Ngoc Anh Mai

Optical flow-based perception, behavior-based control, and topological path planning for mobile robots using fuzzy logic concepts

TUD*press*

Ngoc Anh Mai

Optical flow-based perception, behavior-based control, and topological path planning for mobile robots using fuzzy logic concepts

TUD*press*

2012

Die vorliegende Arbeit wurde am 04. Juni 2012 an der Fakultät Elektrotechnik und Informationstechnik der Technischen Universität Dresden als Dissertation eingereicht und am 31. August 2012 erfolgreich verteidigt.

Gutachter:
Prof. Dr. techn. K. Janschek (TU Dresden)
Prof. Dr. Alfonso G. Cerezo (Univ. of Málaga)

Vorsitzender der Promotionskommission:
Prof. Dr.-Ing. habil. Dipl.-Math. Röbenack (TU Dresden)

Bibliografische Information der Deutschen Nationalbibliothek
Die Deutsche Nationalbibliothek verzeichnet diese Publikation in der Deutschen Nationalbibliografie; detaillierte bibliografische Daten sind im Internet über http://dnb.d-nb.de abrufbar.

Bibliographic information published by the Deutsche Nationalbibliothek
The Deutsche Nationalbibliothek lists this publication in the Deutsche Nationalbibliografie; detailed bibliographic data are available in the Internet at
http://dnb.d-nb.de.

ISBN 978-3-942710-87-9

Verlag der Wissenschaften GmbH
Bergstr. 70 | D-01069 Dresden
Tel.: 0351/47 96 97 20 | Fax: 0351/47 96 08 19
http://www.tudpress.de

Gesetzt vom Autor.
Printed in Germany.

ABSTRACT

OPTICAL FLOW-BASED PERCEPTION, BEHAVIOR-BASED CONTROL, AND TOPOLOGICAL PATH PLANNING FOR MOBILE ROBOTS USING FUZZY LOGIC CONCEPTS

Recently, mobile robots with visual perception working in dynamic environments have been extensively investigated because this method of perception offers a large amount of environmental information. Optical flow perception is an important class of visual perception because it offers powerful perception methods and it offers both egomotion and structure from motion estimation. Especially advantageous is the fact that optical flow perception does not require a priori knowledge of the working environment and can work with minimum hardware, i.e. a mono-camera as the main navigation sensor.

In this thesis, a new approach of optical flow-based perception through qualitative interpretations is developed. Compared to the classical metric approaches for optical flow perception, this approach uses much simpler arithmetic and requires less computation time because of the use of qualitative optical flow interpretations. The qualitative optical flow interpretations provide mobile robots with visual perception a more detailed image of their 3D working environment, e.g. obstacle positions and indoor object types. By using fuzzy logic for the interpretations, the optical flow perception becomes simple and intelligent in a bio-inspired manner and moreover gains robustness under noisy conditions in the working environment.

On the other hand, this thesis develops a generic modular structure of a behavior-based control system with three clearly separate modules for perception, motion control, and path planning. These modules are connected by simple IO interfaces. The system concept is independent of the specific type of perception. The designed behaviors are functionally classified into two separated modules, concerning collision-free motion control and goal-oriented path planning. The hierarchical organization of these behaviors makes the operation of the control system more efficient and enables an easy adjustment of behaviors. Some of the behaviors use fuzzy logic concepts, which result in flexible and smooth robotic motion.

Furthermore a new scheme for topological path planning in combination with fuzzy-based behaviors is developed for the goal-oriented navigation of a mobile robot. This combination allows a mobile robot to perform topological path planning in a real environment without metric information regarding its global and local positions. This enables an easy adjustment of topological path planning for different sensor perceptions or landmarks by just changing the topological map data.

The performance of the optical flow-based perception embedded in the behavior-based control system with the topological path planning has been successfully tested through experiments in a real environment under most realistic conditions including relevant noise effects, e.g. unfavorable lightning conditions, non-standard objects, image processing limitations, image noise, etc.

KURZFASSUNG

OPTISCHER FLUSS-BASIERTE PERZEPTION, VERHALTENSBASIERTE STEUERUNG UND TOPOLOGISCHE PFADPLANUNG FÜR MOBILE ROBOTER UNTER NUTZUNG VON FUZZY-LOGIK

Heutzutage werden mobile Roboter zunehmend mit Kameras ausgestattet, da diese eine Vielzahl von Informationen über die Umgebung bereitstellen. Die Perzeption mit Hilfe des optischen Flusses ist eine wichtige Methode der Bildverarbeitung, da sie eine leistungsfähige Umgebungserfassung und die Nachahmung biologisch-inspirierter Prozesse erlaubt. Dabei können sowohl Informationen zur Eigenbewegung als auch Daten über die Struktur der Umgebung gewonnen werden. Besonders vorteilhaft ist hierbei einerseits die Tatsache, dass keinerlei a-priori-Informationen über die Umwelt benötigt werden und anderseits die geringen Hardwareansprüche von Kamerasystemen. So kann beispielsweise eine einfache Monokamera als Hauptsensor zur Navigation für den mobilen Roboter verwendet werden.

In der vorliegenden Arbeit wird ein neuer Ansatz zur optischen Fluss basierten Perzeption mittels qualitativer Interpretation entwickelt. Verglichen mit klassischen metrischen Methoden, arbeitet der vorgestellte Ansatz dabei mit einer simpleren Arithmetik und benötigt weniger Rechenzeit. Die qualitative Verarbeitung des optischen Flusses bietet dem Roboter ein detaillierteres Bild der dreidimensionalen Arbeitsumgebung. So können beispielsweise Hindernispositionen ermittelt und Objekttypen im Innenraum erfasst werden. Durch die Verwendung von Fuzzy-Logik bei der Interpretation der visuellen Information gestaltet sich die Umgebungserfassung mit Hilfe des optischen Flusses sehr einfach und erlaubt eine bio-inspirierte intelligente Entscheidungsfindung, die auch robust gegenüber realen gestörten Umgebungsbedingungen ist.

Weiterhin wird in der vorliegenden Arbeit eine generische modulare Struktur für eine verhaltensbasierte Steuerung mit drei klar getrennten Modulen für Perzeption, Bewegungssteuerung und Pfadplanung vorgestellt. Diese Module werden über einfache Schnittstellen miteinander verbunden. Dadurch ist das entstandene System auch auf andere Perzeptionsmethoden mobiler Roboter anwendbar. Die realisierten Verhaltensmuster werden dabei funktionsorientiert in zwei Module eingeordnet: Ein Modul sichert hierbei die kollisionsfreie Bewegungssteuerung, ein weiteres realisiert die zielorientierte Pfadplanung. Die hierarchische Organisation dieser Verhaltensmuster ermöglicht ein effizientes und einfaches Vorgehen bei der Modifikation der hinterlegten Eigenschaften. Dabei nutzen manche dieser Verhaltensmuster wiederum Konzepte der Fuzzy-Logik, um die Roboterbewegung so flexibel und leichtgängig zu realisieren, wie es bei biologischen Systemen der Fall ist.

Für die zielorientierte Navigation eines mobilen Roboters wurde in einem dritten Schwerpunkt eine neue Methode für die topologische Pfadplanung in Kombination mit Fuzzy-Logik-basierten Verhalten entwickelt. Diese Kombination ermöglicht dem Roboter die topologische Pfadplanung in einer realen Umgebung ohne jegliche Verwendung von metrischen Informationen in Bezug auf seine Position und Orientierung. Dadurch kann die Pfadplanung durch einfache Modifikationen der topologischen Kartendaten für verschiedene Perzeptionssensoren oder Landmarkenrepräsentationen angepasst werden.

Die Leistungsfähigkeit der Perzeption mittels des optischen Flusses innerhalb der verhaltensbasierten Steuerung zusammen mit der topologischen Pfadplanung wird anhand von Experimenten mit einem mobilen Roboter in einer realen Umgebung gezeigt. Dabei werden auch unterschiedlichste Bedingungen, wie sich ändernden Lichtverhältnissen, unbekannten Objekten, Einschränkungen bei der Bildverarbeitung sowie Bildrauschen berücksichtigt.

Acknowledgment

This work has greatly benefited from the support of many people over the past years. I would like to express my deep gratitude at its completion, especially to those mentioned below.

To begin, I would like to thank my supervisor, Prof. Klaus Janschek, for supporting and guiding me from the start point of my PhD, where I was a fish out of water, to the end point of my completion. Especially, he helped me to believe and pursue my ideas.

Thank you to my best friends Matthias Hennig and Sylvia Horn for helping me to develop working and study skills, and giving me helpful advices paving the way for my accomplishment. They pushed me to fall in line with the interesting life in Germany. Additionally, Matthias's intelligence and humor make my job and life more comfortable.

To Martin Beck, who is in the heaven, thanks for filling me a large number of knowledge about optical flow and image processing.

My colleagues at the IfA have made my time enjoyable and rewarding. I thank you for consulting and supporting me in the field of robotics. Especially, I must thank Frank Schnitzer for his valued friendship.

Finally, I thank my lovely wife Ly, who supports the constant love and happy life. Especially, I must thank her for giving me the most beautiful gift, my daughter. To my parents, thank you for supporting me over the years and providing me a lifetime of thoughtful care.

– For Ly
and
in memory of Martin Beck –

Contents

Symbols and Abbreviations

Abbreviations

AMR	Autonomous Mobile Robot
Aux.	Auxiliary
Cam	Camera
DOA	Degree Of Applicability
DOF	Degrees Of Freedom
FIS	Fuzzy Inference System
FOE	Focus Of Expansion
FOV	Field Of View
GPU	Graphics Processing Unit
IfA	Institut für Automatisierungstechnik
IO	Input Output
LOS	Line Of Sight
MOFIS	MOtion control Fuzzy Inference System
OF	Optical Flow
OFFIS	Optical Flow Fuzzy Inference System
PAPFIS	PAth Planning Fuzzy Inference System
SLAM	Simultaneous Localization And Mapping
stdv	Standard deviation of a data set
TIO	Typical Indoor Object
TPO	Two-Plane built Object

Coordinate frames

{B}	Body frame of the mobile robot
{C}	Compass frame
{I}	Inertial frame

Functions

f_{tilt}	Multiplication of a pair of tilts
hplane	Identifies a horizontal plane
max	Computes the maximum value of a set of data
mean	Computes the mean value of a set of data
min	Computes the minimum value of a set of data
plane	Identifies a plane
tilt	Computes tilt of a plane
vplane	Identifies a vertical plane

Matrices and vectors

Ad	Adjacency matrix of topological map
$\mathbf{dir}_O$	vector of obstacle directions
$\mathbf{d}_O$	vector of obstacle distances
$\mathbf{d}_L$	vector of landmark distances
$\mathbf{d}_{\Omega^{II}}$	matrix of fuzzy distances based on layer-2 quadrants
fineShape	matrix of fine shape of 3D structure
obstacleDirection	matrix of obstacle directions
obstacleDistance	matrix of obstacle distances
roughShape	matrix of rough shape of 3D structure

tiltL2	matrix of layer-2 quadrants
tiltL3	matrix of layer-3 quadrants
TPOdistance	matrix of TPO distances
v	vector of velocity
$\mathbf{\Omega}^{\mathrm{I}}$	matrix of layer-1 quadrants
$\mathbf{\Omega}^{\mathrm{II}}$	matrix of layer-2 quadrants
$\mathbf{\Omega}^{\mathrm{III}}$	matrix of layer-3 quadrants

Variables

u	Magnitude of a optical flow vector mapping on U axis
υ	Magnitude of a optical flow vector mapping on V axis
α_{G}	Goal-oriented angle
δ_{DL}	Dangerous radius around a deadlock
δ_{G}	Safe distance to a goal
act_{G}	Motion action at a local goal
A_{i}	Indoor situation related to TPO
d_{F}	Fuzzy distance to object
d_{max}	Maximum distance
d_{min}	Minimum distance
Dp_{G}	Depth of goal
d_{L}	Distance to landmark
d_{O}	Distance to obstacle
d_{S}	Distance to object
endP	End point
GdAct	Guidance action for robot motion
lGoal	Local goal
Goto	Action motion for going to goal
Lmk	Landmark type
pth	Path to target
Q_i	Quadrant *i* of an optical flow field
rmn	Ratio of mean
rsd	Ratio of standard deviation
set_{G}	Command of setting goal
startP	Start point
$state_{\mathrm{G}}$	Found-goal state
S_{L}	Landmark type of an observed object
Through	Action motion for going through goal
v	Linear velocity
YA	Yaw angle
ε	Direction offset
θ_{G}	Goal direction
θ_{R}	Robot direction
μ	Mean
σ	Standard deviation
Σ_{E}	Set of edge attributes
Σ_{N}	Set of node attributes
ω	Angular velocity

List of special figures

Chapter 1
Introduction

1.1. Motivation

There are many ways for mobile robots to deal with navigation tasks in unknown and changing environments. Over recent decades, a large number of robotic projects have investigated the potential of fuzzy information processing schemes as an alternative to standard metric information processing for mobile robot control, because these schemes require less precision and less precise information from the working environment. Additionally, it is expected that simpler algorithms and faster processing, e.g. less computational effort, can be applied to these schemes.

The motion control of indoor mobile robots is an important task because it strongly relates to actual problems of service robots and needs highly flexible and robust control algorithms. For this reason, behavior-based control has emerged as a reasonable candidate because it offers the required flexibility and compatibility with fuzzy information processing to deal with dynamic environments.

A lot of applications of behavior-based control for mobile robots have been developed in various working environments such as industrial transportation (MB385), medication delivery (TUG), nursing (Pollack et al. 2002), and assistance (Tsai et al. 2009) for the elderly. In those applications the robots work effectively concerning the variations and uncertainties of the environments.

Furthermore, mobile robots with powerful visual perception have recently been developed because visual perception offers robots a large amount of environmental information. Optical flow perception is an important class of visual perception because it forcefully offers perception methods in biological systems, e.g. the insects' systems, and offers egomotion and structure from motion. Additionally, optical flow perception does not require a priori knowledge of the working environment and can work with minimum hardware, i.e. a mono-camera as the main navigation sensor.

The classical approach for solving the navigation problem of "…determining and maintaining a course or trajectory to a goal location" (Franz and Mallot, 2000) is based on metric reconstruction of the robot's pose (position and orientation), metric modeling of the environment, metric path planning, and locomotion control. These engineering solutions to a basic biological problem demonstrate many considerable differences to biological solutions. Therefore, in the past decades a lot of effort has been invested in copying biological solutions in man-built robotic machines, see (Franz and Mallot, 2000) for a rather comprehensive review of biomimetic robotic navigation engineering solutions.

Many approaches, e.g. (Coombs and Roberts, 1993), (Santos-Victor, 1993), (Tian et al, 1996), (Dev et al. 1997), (Zucchelli et al. 2002), (Zufferey and Floreano, 2006), (Tchernykh et al. 2006), succeed in solving the navigation problem of mobile robots based on metric a priori information about the working environment and on metric environmental information extracted from optical flow, resulting in general in rather complex computations.

In contrary to such approaches this work investigates optical flow-based perception and behavior-based control for mobile robots using topological path planning and fuzzy logic concepts to deal with navigation tasks in dynamic indoor environments with less metric information and more moderate computational efforts.

1.2. State of the art

In the last few decades, the development of indoor service robots concerning transportation missions has been increasing forcefully in many different application areas. There are many well-known indoor service robots for transportation such as HelpMate (Krishnamurthy and Evans, 1992) and (Evans, 1994), i-Merc (Carreira, 2006 and 2007), serving inside hospitals and health care centers, and the ARIADNE robot team (Pauly et al. 2002), delivering food, drinks, or mail in official areas.

1.2.1. Visual mobile robots

Among today's service robots, visual mobile robots have been extensively investigated because of their remarkable capability to retrieve the structural information of environments and to react very quickly to uncertain and changing environments based on the perceptional simplicity, which needs at least a mono-camera as the main navigation sensor. Vision-based methods of perception play an important role in the self-localization of autonomous service mobile robots operating in indoor environments.

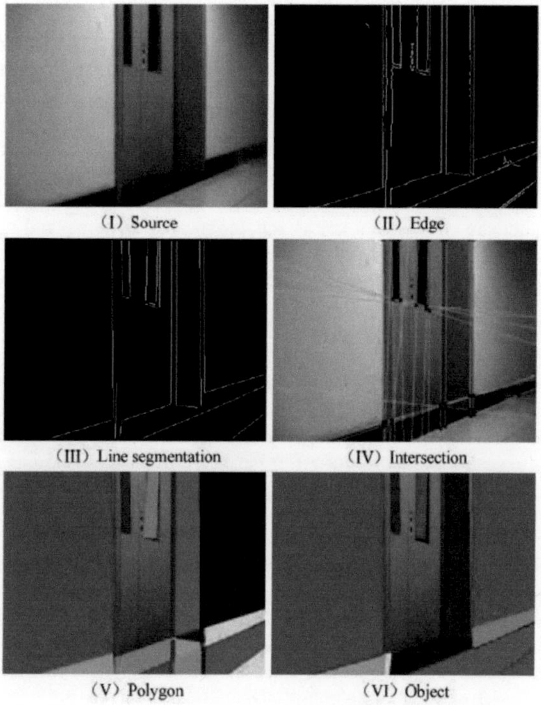

Fig. 1.1. Vision-based object segmentation in a corridor scene using polygon homogenous color test and merge test (Chai et al. 2011).

To determine the position and orientation of robots, some of the vision-based mobile robots basically rely on matching observed specific features of natural objects in indoor environments with expected ones. For example, (Jeong and Lee, 2006) proposed a vision-based SLAM (Simultaneous Localization And Mapping) technique using both line and corner features as landmarks in the scene to localize and reconstruct 3D line and corner landmarks at the same time using an extended Kalman filter-based framework. The central idea of the proposed algorithm is based on a corner matching between camera images and a stored 3D Corner map, between line projections from camera images, and a stored 3D Line map to find the best matching candidates for robot poses.

In a more recent work (Chai et al. 2011) a vision-based object segmentation technique that extracts straight-line features from the indoor scene was proposed. The key concept of the proposed method is combining straight lines to form many convex polygons. Polygons with homogenous surface color were processed to detect the forms of indoor objects, see Fig.1.1.

The approaches mentioned above are referred to in order to determine absolute robot position and orientation, in other words absolute localization, because such navigation systems must construct a match between observations and expectations derived from a stored database. Obviously ambiguities in such localization may happen with uncertainties, which are associated with observations that match multiple expectations. For this reason, several methods for resolving ambiguities have been proposed, such as partially observable Markov models (Simmons and Koenig, 1995), Monte Carlo localization (Dellaert et al. 1999), and Markov localization (Thrun, 2000).

In contrast to the absolute localization methods, incremental localization methods strongly depend upon the geometry or metric information of environments for self-localization. For example, a Honda ASIMO humanoid robot employed "an overhead camera to compute the position of the robot, the desired goal location, indicated by a marker, and all obstacles" (Michel et al. 2005). Under the same assumption, in (Andersen et al. 1995), an autonomous guided vehicle used a camera mounted on the ceiling to precisely predict its positions.

The main concepts of incremental localization methods are the use of geometrical representation of space, the use of propagation of positional uncertainty through commanded motions, and the projection of the robot's positional uncertainty into a camera image.

Generally, the vision-based localization methods commonly rely on metric information and complex computations.

1.2.2. Optical flow-based navigation

In recent years, many research groups have attempted to develop vision-based mobile robots with biological navigation behaviors by copying biological solutions in man-built robotic machines. One of the biomimetic robotic navigation engineering solutions that has been investigated deeply is the *optical flow* navigation. According to (Horn and Schunck, 1981), optical flow is the distribution of apparent velocities of movement of brightness patterns in an image, and it arises from the relative motion of objects and an observer (camera, robot).

Optical flow navigation is based on the motion of image points in camera images recorded from a moving camera. It employs so-called "motion stereo" from mono-camera images, and it has been investigated very extensively for robotic vehicle navigation applications in the last few decades.

The optical flow field allows for the reconstruction of a 3D structure from 2D images. The classical optical flow navigation solutions provide pure metric models for both egomotion and structure-from-motion determination (Pranzdny, 1980), (Tian et al. 1996), (Zucchelli, 2002), (Janschek et al. 2006), (Tchernykh et al. 2006).

(Coombs and Roberts, 1993) developed an optical flow-based system that mimics the visual behaviors of bees by balancing the image flow for flight in a corridor or evaluating the maximal flow for nearest obstacle detection. In the same year, (Santos-Victor et al. 1993) published a divergent stereo approach, which also mimics the behaviors of bees with smoothed-centering between averaged object distances. The main idea of the divergent stereo approach is that differential estimation of the optical flow measured by two cameras mounted on a robot, named Robee, makes a reflex-type motion control which mimics the reflex-centering of a bee. Moreover, this robot used motion-derived features, such as "time-to-crash" rather than depth information.

A few years later, (Zufferey and Floreano, 2006) developed autonomous micro-flyers, shown in Fig. 1.2, capable of navigating within houses or small indoor environments by using vision as the principal source of information. They used optical flow divergence processing for obstacle avoidance based on severe weight and energy constraints.

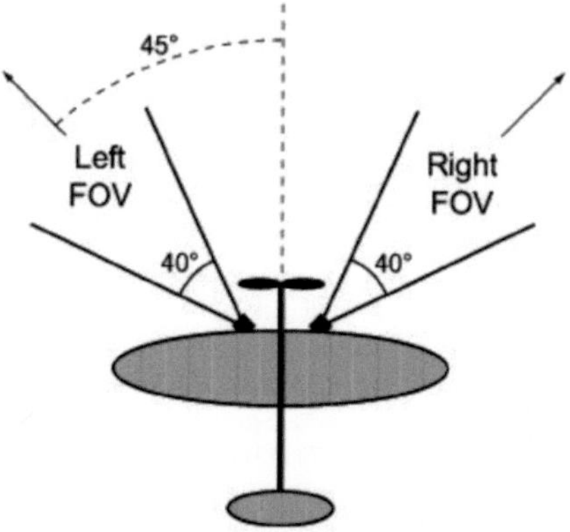

Fig. 1.2. Top view of airplane with two 1-D cameras in (Zufferey and Floreano, 2006)

Generally, the discussed optical flow-based solutions use the spatial derivatives of optical flow field for control strategies. However, despite their bee mimicry, these solutions disregard object recognition and 3D information from the scene.

Choosing a different path, (Dev et al. 1997) utilized the temporal derivative of the optical flow field for wall-following navigation. The key aspect of the approach is the extraction of depth information from optical flow. In order to reduce noise within the optical flow, the authors did not use the entire optical flow field but instead took the average of fine optical flow at the corners or borders of optical flow images.

(Naito et al. 2007) also used depth information of the optical flow, estimated 3D information by including the depth and height of obstacles based on optical flow extracted along the edges of images. However, the complexity of this approach is in the detecting corners in images using the Harris Corner Detector (Harris and Stephens, 1988).

Other authors, e.g. (Enkelmann, 1991) and (Sull and Sridhar, 1999), proposed model-based approaches using optical flow to detect not only obstacles, but also dominant planes. For example, (Sull and Sridhar, 1999) stated "by the estimation of spatial and temporal residual disparity, the disparity values can be controlled up to a certain degree. The disparity analysis also shows that the location of FOE (Focus of Expansion) is very important since obstacles cannot be detected near FOE if temporal disparities are the only source of data available. For this area, the spatial disparities should be used."

(Ohnishi and Imiya, 2006) suggested a non-model-based approach to detect a dominant plane. The key idea of the approach is the segmentation of an image sequence using optical flow in the detection of a dominant plane in an image sequence. They defined that "the dominant plane is the planar area which occupies the largest domain in the image observed by a camera," and assumed that "the dominant plane in front of the robot is the largest area on the floor on which the robot moves," and "the dominant plane occupies over half of the image at the initial frame."

Fig. 1.3 illustrates the procedure for dominant plane detection proposed in (Ohnishi and Imiya, 2006). The procedure can be simply explained as follows: first, randomly select three points in the optical flow. Second, compute the affine coefficients at the selected area. Third, estimate planar flow from the affine coefficients. Finally, match the computed optical flow and estimated planar flow to detect a dominant plane, which must occupy more than 50% of the image.

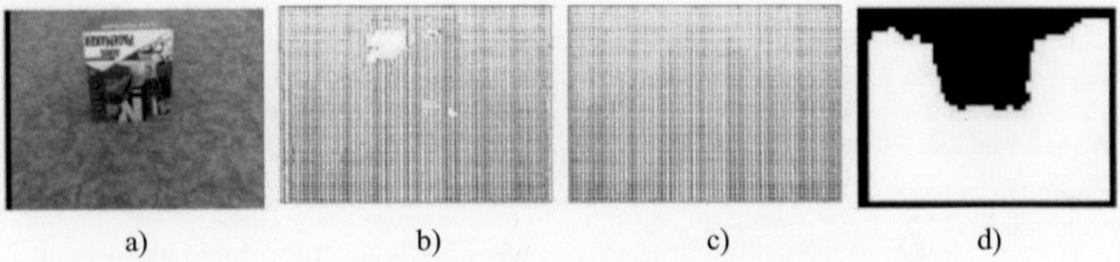

Fig. 1.3. Procedure for dominant plane detection in (Ohnishi and Imiya, 2006):
a) original image sequence, b) computed optical flow,
c) estimated planar flow, d) detected dominant plane (white area)

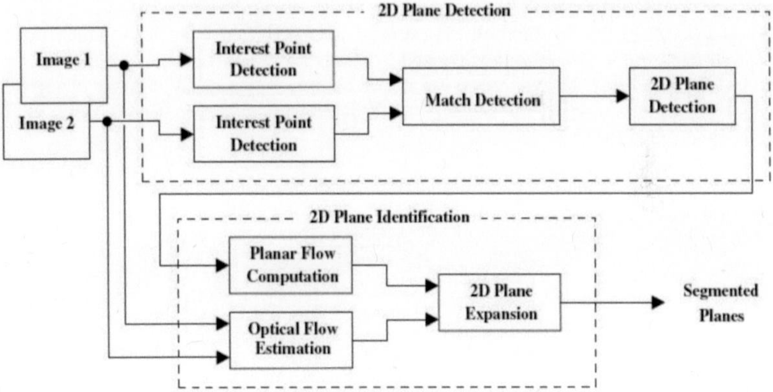

Fig. 1.4. Block diagram of the plane segmentation system proposed in (Aires and Mediros, 2010)

In a more recent work, (Aires and Mediros, 2010) implemented a plane segmentation system based on affine homography and optical flow. According to (Aires and Mediros, 2010), the 2D plane detection, performed by the upper block in Fig. 1.4, consists of a clustering scheme based on the computation of homographies using the model affine. The detection procedure provides multiple planar regions in the 2D image space. The 2D plane identification, performed by the lower block in Fig. 1.4, uses optical flow to classify the planar regions into two categories, ground floor and obstacles, in that the ground floor is the predominant plane in the image, and that obstacles are represented by small secondary planes.

To detect dominant planes, namely floors, the mentioned approaches rely on the number of computations in the image points belonging to each segmented plane. Identifying obstacles and ground planes consumes a large part of computing time. Additionally, the approaches can only handle the problem of close obstacles due to the small planar ratio of the dominant planes.

Although most of the optical flow-based solutions listed above are, general speaking, inherently biomimetic in their approach, they are based on more or less *classical quantitative decision* making implementations, require *complex computations*, and support *poor information* about 3D structures.

1.2.3. Behavior-based control using fuzzy logic

A large number of mobile robots work in unknown environments with many uncertainties. To cope with these uncertainties, behavior-based robotic systems have recently emerged as one of the preferred solutions. Behavior-based robotic systems optimally serve when "the real world cannot be accurately characterized or modeled" (Arkin, 1999), and they can "deal with various situations without being provided models or environment maps" (Watanabe et al. 1992).

Being inspired by the biological world, in the past decades many behavior-based robotics researchers have tried to answer dozens of questions, such as, "How does animal behavior relate to perception and action?" "What can animal behavior offer robotics?" "How can we make a robot behave like an animal?" "How are multiple behaviors controlled and coordinated?" These researchers argue that "there is much that *can* be gained for robotics through the study of neuroscience, psychology, and ethology" (Arkin, 1999).

Since the mid-1980s, behavior-based architecture has emerged as an alternative to traditional approaches for designing autonomous mobile robots. (Brooks, 1986) proposed the subsumption architecture in order to control mobile robots. In this architecture, a complex navigation task can be decomposed into several well-specified behaviors, which operate in parallel to each other. An arbitrator is used to resolve any conflicts arising from competing behaviors to determine the control action of the complete system. A number of the existing behavior arbitrators adopted the strategy of High-Priority-Take-All (Beom and Cho, 1995) and (Yung and Ye, 1996).

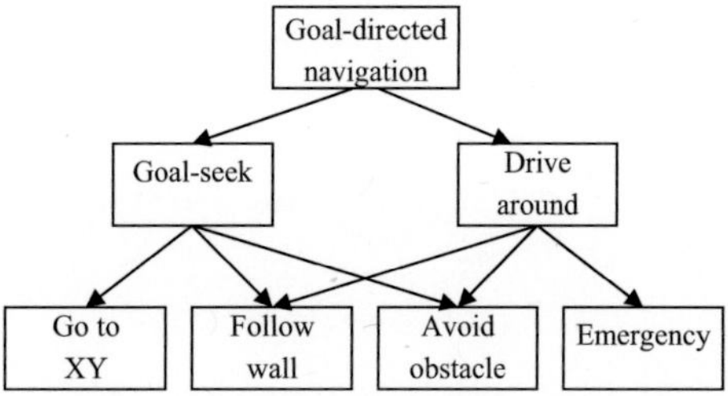

Fig. 1.5. Hierarchical architecture of intelligent behavior system (Tunstel, 1996)

(Tunstel, 1996) developed a behavior-based robot navigation control system with a hierarchical structure of fuzzy-based behaviors shown in Fig. 1.5. The general robot behavior is decomposed into a bottom-up hierarchy of increased behavioral complexity in which activity at a given level is dependent upon the behaviors on the levels below it. A collection of primitive behaviors resides at the lowest level which is referred to as the primitive level. Each of these primitive behaviors is simple, self-contained, and designed to carry out a single control policy or goal. Noticeably, the low-level navigation behaviors are implemented as fuzzy logic controllers, while the higher-level coordination behaviors work as fuzzy decision systems.

The key idea of Tunstel's approach is that the fuzzy-based hierarchical architecture enables the distribution of intelligence among behaviors by using the concept of *degree-of-applicability* (DOA) for weighted decision-making coordination. The coordination executes behavior selection by the control policy: "behaviors with maximal applicability can be said to dominate, while behaviors with partial applicability can be said to inhibit" (Tunstel, 1996).

(Seraji and Howard, 2002) proposed a behavior-based strategy for visual robot navigation to achieve terrain assessment, using also fuzzy logic rules to adjust the weighting factors of behaviors. The central features of the approach concern the following aspects: processing camera images to extract three terrain characteristics, including roughness, slope, and discontinuity; creating three fuzzy-based behaviors comprising of traverse-terrain behavior, avoid-obstacle behavior, and seek-goal behavior; automatically adjusting the weighting factors of the behaviors based on the situational context of the robot; integrating the three behaviors through the weighting factors to generate a final motion command for the robot.

In comparison with (Tunstel, 1996), the fusion behavior in (Seraji and Howard, 2002) is smoother due to all three primitive behaviors being included in the final behavior by applying the Center-of-Largest-Area (COLA) defuzzification method. However, it is difficult to represent the COLA defuzzification method formally. Moreover, the smoothness of the final behavior happens mainly between the traverse-terrain and avoid-obstacle behaviors, because the weighting factor of the seek-goal behavior is nominal at all times.

In more recent works (Bao et al. 2009) and (Wang et al. 2011), the incorporation of fuzzy logic control and *priority-based* behavior control in mobile robot navigation in unknown environments has been implemented. The authors created four fuzzy-based behaviors, those being goal-seeking, obstacle avoidance, wall-following, and deadlock-disarming. However, the final integration of the behaviors does not rely on fuzzy rule bases, but rather on priority evaluation shown in Fig. 1.6. That means the referred works deal with problems, which contain complex situations and require more than one behavior at a moment, in a difficult and inefficient way.

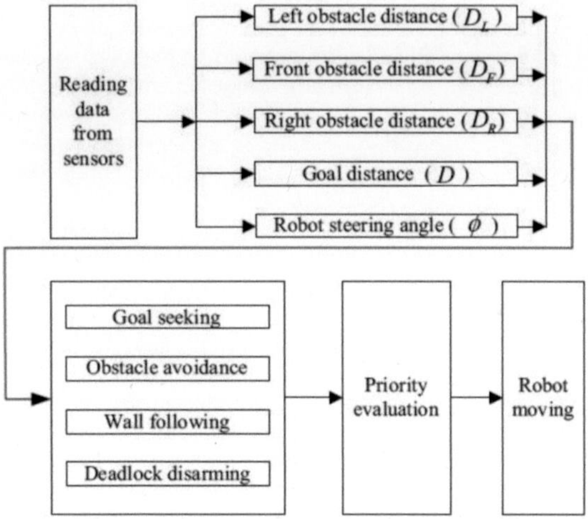

Fig. 1.6. Behavior-based hierarchical fuzzy controller architecture proposed in (Wang et al. 2011)

(El-Teleity et al. 2011) introduced a behavior-based control system using fuzzy logic for the navigation of the Khepera I mobile robot. The robot has three fuzzy-based behaviors for reaching targets, avoiding obstacles, and escaping traps. Particularly, a fuzzy supervisor, illustrated in Fig. 1.7, has been applied for the fusion of all behaviors. The fuzzy supervisor determines the execution priority of all elementary behaviors according to five selection inputs, including the distance between robot and target, the left and right proximity sensors, the state of the robot representing the "normalized absolute value of the alignment error," and the initial wall-follow state representing "the polarity of the angle at the moment of firing one of wall-follow behaviors (left and right walls)" (El-Teleity et al. 2011).

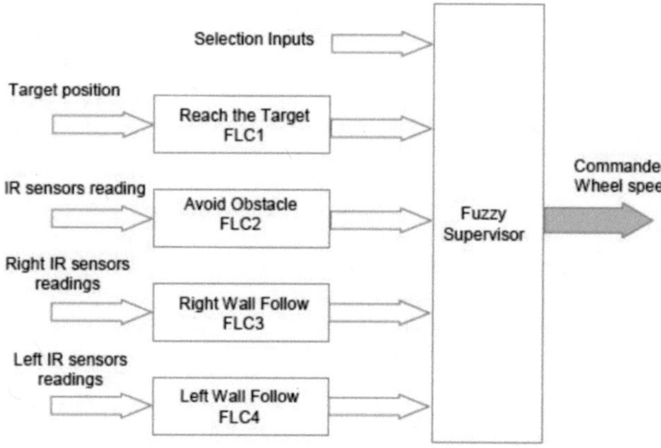

Fig. 1.7. Hierarchical behavior-based controller scheme (El-Teleity et al. 2011)

To summarize, although behavior-based controllers using fuzzy logic have been widely applied, the *smooth fusion* of all behaviors needs to be improved for complex environments requiring multiple behaviors simultaneously. Moreover, the *complexity* and *computational cost* of fuzzy-based fusions should be reduced. Furthermore, a *generic modular hierarchical architecture* for behavior-based control systems is not known so far, which would allow for the easy adaption of new navigation tasks by adding or changing behavior modules.

1.2.4. Topological path planning

Path planning is a fundamental problem in mobile robotics. The objective of path planning is to find a collision free path that leads a robot from a start point to a target in the working environment, which contains static and/or dynamic obstacles.

In the mobile robot systems, path planning may be classified as local (sensor-based) and global (model-based) path planning.

On one hand, global path planning aims to deal with the issues of long-term planning concerned with finding of the best path to the target based on foreknown information of the working environment. In other words, the "global path planning requires the environment to be completely known and the terrain should be static" (Sedigh et al. 2004).

On the other hand, local path planning has to solve problems of short-term planning involved in finding of local goals and avoiding unknown obstacles. This means the "local path planning is done while the robot is moving" (Sedigh et al. 2004). In other words, in local path planning, a given path may be varied because the working environment may change.

Much research on path planning has been done on dynamic environments (Fraichard, 1993), (Mechlih, 1993), (Jia and Wang, 2009) and (Tang et al. 2010). However, they use information of metric coordinates (Fraichard, 1993), (Mechlih, 1993), and/or of time with "the time history of a continuous sequence of configurations between the current configuration of the robot and its goal configuration" (Jia and Wang, 2009).

Topological path planning also called qualitative path planning is one kind of path planning using a topological map to represent the working environment. The topological map illustrates the working environment by a graph structure where graph nodes define the landmarks and graph edges define procedural information for the movement between the nodes (Tarutoko et al. 2006). Several graph-based search algorithms for topological path planning are introduced in (Russell and Norvig, 2003).

There are lots of known approaches for topological path planning: some of them use metric information, e.g. (Thrun, 1998) uses a grid-based map concerning metric dimensions, (Abraham et al. 2009) uses the metric coordinates of environments, (Chen et al. 2011) uses the metric distance information between the nodes of a topological map; some others use non-metric information, e.g. (Lin et al. 2004), (Lee et al. 2006) and (ElKaissi et al. 2007). Their solutions, however, have been developed for particular sensors, e.g. (Lee et al. 2006) uses sonar sensors, (Lin et al. 2004) uses WEB camera, and (ElKaissi et al. 2007) uses a space and edge laser range finder.

So far a general scheme for topological path planning which is independent of the perceptional source is not known to the author.

1.3. Objectives of the thesis and achieved contributions

Stemming from the discussed motivations, the basic objectives of this doctoral thesis are the following:

O1: Development of a robust qualitative interpretation approach for optical flow-based perception for indoor mobile robots.

Achievements: The optical flow-based perception incorporates a new concept for qualitative interpretation of regionally averaged optical flow for structure from motion and object recognition. The qualitative interpretations use fuzzy logic to make optical flow-based perception more reliable and robust under the uncertainty and noise of indoor environments. The algorithms of the qualitative interpretations must be simple and fast.

O2: Design of a generic modular structure of behavior-based control system containing perception, motion control, and path planning modules using fuzzy logic.

Achievements: The behavior-based control system is designed as a generic modular structure containing clearly separated and self-standing modules for perception, motion control and path planning. The first module, the perception module, provides sensor-based information about the surrounding environment. The two last modules hold behaviors organized in a transparent hierarchy to improve the intelligent ability of multi-behavior management. Additionally, the behavior-based controller applies fuzzy logic for several behaviors and behavioral fusion to make robot motion smoother in complex environments. Noticeably, the modules of the generic modular structure are able to be replaced regardless of the inner algorithms because of clear and simple IO interfaces.

O3: Design of a generic scheme for topological path planning independent of the perceptional source.

Achievements: A generalized scheme for topological path planning has been developed which is combined with a behavior-based control subsystem using fuzzy logic to navigate a mobile robot to a given target. This combination makes the goal-oriented navigation flexible and robust with respect to complex and dynamic environments though lacking metric information.

It is further noticed that the achievements of objectives O2 and O3 are generally applicable to different mobile robot platforms and they are independent of the perceptional source.

1.4. Organization of the thesis

Subsequent to this chapter, the thesis is organized as follows:

Chapter 2: Concept of Behavior-based Motion Control System

To set the fundamentals for designing the navigation system, this chapter provides a brief introduction of design principles for behavior-based systems. Subsequently, the functional architecture of the behavior-based controller is given. The controller comprises three main modules, including the OFFIS module responsible for optical flow perception using Fuzzy Inference Systems (FIS), the MOFIS module undertaking motion control and applying FIS for behaviors of motion control, and the PAPFIS module working as a path planner and taking responsibility for the computation of the fuzzy-based behaviors of path planning.

In the last part of the chapter, a general description of the basic concepts of the OFFIS, MOFIS, and PAPFIS modules is provided. Further details for these modules will be described in the later chapters.

This chapter discusses the fundamental results of the *objective O2*.

Chapter 3: Fuzzy-based Optical flow Perception (OFFIS)

Firstly, the basics of optical flow and the basic concepts of optical flow image interpretation involving topological structure from motion, hierarchical segmentation, and two-plane built objects are presented briefly. After that the statistical optical flow metrics that play an important role in the qualitative interpretations of this work are discussed in detail to try and deduce why the mean value of the optical flow vector subfields have been chosen as sole metrics.

Subsequently, the algorithms for structure from motion are proposed. The proposed algorithms are used for plane detection, tilt-angle calculation, and structure reconstruction.

Finally, fuzzy-based algorithms for qualitative interpretations are described in detail. The results of the qualitative interpretations are auxiliary information about obstacle direction, object type, and fuzzy distances. Additionally, a novel scheme of outlier detection based on qualitative detection of optical flow outliers on regionally averaged optical flow patterns is represented. This filter solves the challenging problem of reducing the noise of optical flow field in real environments.

This chapter unveils the results of the *objective O1*.

Chapter 4: Fuzzy-based Motion Control (MOFIS)

To establish the basics of behavior-based control, this chapter first introduces a description of mobile robot kinematics and the basics of behavior-based robotics. After that, the MOFIS primitive and composite behaviors, including fuzzy logic, are described in detail. At the end of this chapter, the selections of MOFIS behaviors and kinematic fusion are depicted. The kinematic fusion is executed by a fuzzy-based cooperative coordination taking local information about obstacles and topological path planning information into account.

This chapter presents the results for the *objective O2* concerning motion control.

Chapter 5: Fuzzy-based Path Planning (PAPFIS)

To install the fundamentals of topological path planning, the general concepts of the topological map are presented at the beginning. Next, the global planner that uses a graph-based search algorithm for the topological path planning and the local planner are illustrated.

After that, this chapter describes the fuzzy-based behaviors of path planning, including primitive and composite behaviors, and the behavior selection of PAPFIS.

Along with the stated objectives, this chapter presents the results of *objective O3* and *objective O2* concerning path planning behaviors.

Chapter 6: Simulation based test and verification

This chapter describes simulation setups, comprising the POV-Ray and Matlab simulation environment, and discusses the simulation results.

The objective of the simulation is to test and verify the proposed algorithms and control system design with various navigation tasks in a closed-loop configuration.

Chapter 7: Experimental test and verification

This chapter presents the experiment setups concerning the experiment environment, hardware, and data connection. The experiment results are then illustrated and discussed.

The objectives of the experiments are to test and verify the proposed algorithms and control system design, similar to the objectives of the simulations, and to check the reliability of the optical flow-based perception and the robustness of the behavior-based control system under real world conditions.

Chapter 8: Discussion and Conclusion

This chapter summarizes the proposed algorithms and system design, according to the results of simulation and experiments. Finally, this chapter addresses the contributions of this work and discusses possible future work.

Chapter 2
Concept of behavior-based motion control system

This chapter addresses the new generic modular hierarchical architecture for a behavior-based control system with clear interfaces for the path planning and motion control systems. This architecture allows for easy adaption of new navigation tasks by adding or changing behavior modules.

In the following, the design principles of behavior-based control systems are introduced. A functional architecture of the behavior-based controllers is given that is based on these design principles. The functional architecture mainly consists of three modules: the OFFIS module- responsible for optical flow (OF) perception using Fuzzy Inference Systems (FIS), the MOFIS module- incorporates behavior-based motion control (MO) using fuzzy logic while taking into account the local information about obstacles and path planning information, and the PAPFIS module- performs path planning (PAP) and executes fuzzy-based behaviors concerned with goal-oriented navigation. Finally, these modules are described in accordance with the functional architecture developed in this chapter.

2.1. Basic principles for designing behavior-based control systems

The following basic principles for designing behavior-based systems have been widely used in behavior-based robotics. These principles, see (Pfeifer, 2005), are listed as follows:

- **Modularity**. The structure of the system has to be established in various modules.
- **Interface**. The input and output (IO) interface of modules must be independent of the change within the internal structure of the modules. The modules of the system can be replaced with sole consideration of IO interfaces.
- **Perception**. The system can acquire information about the environment through its own sensory systems as a result of its interaction with the real world.
- **Representation**. The robot system has the ability to represent the environment based on internal data, such as a topological map.
- **Planning ability**. The system planner might confront robots with paths and motion actions that will allow them to accomplish navigation tasks in a self-sufficient manner.
- **Hierarchical organization**. Behaviors are planned and executed on several levels. Decisions involved in the tasks or objectives of planning are at the highest level.
- **Parallelism**. Behaviors on the same hierarchical level are realized concurrently.
- **Expendability**. The control system allows for the addition of new behaviors without having to modifying the current ones.

2.2. Functional architecture of the behavior-based controller

Based on the design principles stated above, the optical flow-based navigation system with the behavior-based robot controller is illustrated in Fig. 2.1.

The optical flow-based navigation system uses a *mono-camera* (Cam) as the main navigation sensor which provides imaging information about the working environment.

The optical flow processor processes the camera images in pairs to extract optical flow vectors that are necessary for obtaining high level image navigation information.

The extracted optical flow contains components from both translational and rotational camera motion. Since only the translational optical flow component contains 3D information (motion stereo), the rotational component must be removed before starting the optical flow processing (Prazny, 1980). This is easily accomplished by using an auxiliary sensor (Aux.), such as a gyroscope or odometry for *angular-rate* information.

The auxiliary sensor is also used to determine the *robot-direction* and odometry necessary for topological path planning, and to compute goal-oriented behaviors.

The behavior-based controller uses the following input data (see. Fig. 2.1): *inertial data* for information measured by auxiliary sensors (from gyroscope), *perceptional data* for information taken from perception (optical flow field), *planning data* for information computed by a path planner (topological map), and *user-defined data* for information given by user (mission objective).

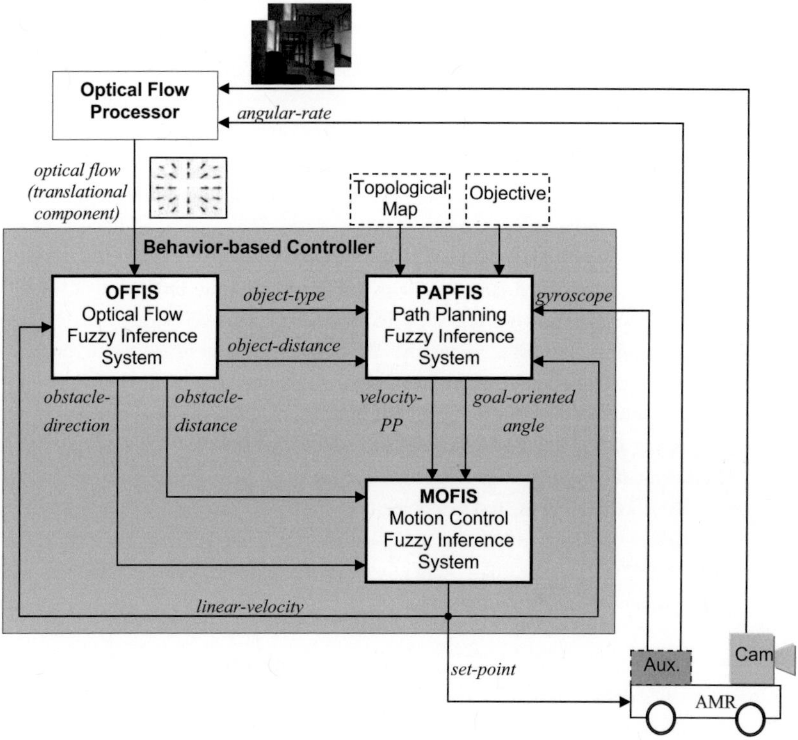

Fig. 2.1. Functional architecture of the optical flow-based navigation system for behavior-based robot control with fuzzy inference systems

The optical flow perception function of the control system is realized by the Optical Flow Fuzzy Inference System (OFFIS). The translational component of optical flow taken from the optical flow processor will be segmented into quadrants to execute an algorithm of qualitative optical flow processing to extract topological 3D information about the working environment.

The OFFIS module delivers perceptional data: local information about landmark including indoor landmark type *object-type* and fuzzy landmark distance *object-distance*, and local information about obstacle including *obstacle-direction* and fuzzy *obstacle-distance*.

The planning function of the control system is executed in the PAth Planning Fuzzy Inference System (PAPFIS). The PAPFIS module carries out topological path planning based on inertial data, e.g. *gyroscope* or *odometry*, perceptional data, e.g. *object-type* and *object-distance*, and user-defined data, e.g. a *topological map* representing the world and the *objective* of a navigation task containing path points (*start point* and *end point*).

The PAPFIS module provides the motion control with planning data: a velocity command *velocity-PP* for goal-oriented navigation to drive the robot as it approaches a local goal, and a *goal-oriented angle* to identify robot direction as it compares to the local goal direction.

The motion control function is carried out by the MOtion control Fuzzy Inference System (MOFIS), which is based on the local obstacle information, and by the planning data of the PAPFIS module. The MOFIS module computes behaviors for collision-free navigation and performs a kinematic fusion of the goal-oriented behavior and collision-free behavior to generate a set-point for locomotion control of the AMR.

The current work mainly concentrates on these three modules of the behavior-based controller: OFFIS, MOFIS, and PAPFIS.

2.3. Modules of the behavior-based controller

2.3.1. OFFIS module

The OFFIS module, illustrated in Fig. 2.2, executes a qualitative optical flow-based perception using fuzzy logic. Through its perception, a mobile robot acquires local information about obstacles and indoor landmarks in the working environment.

The input of OFFIS is the translational component of optical flow vectors taken from the optical flow processor. It is assumed that the optical flow processor performs a basic preprocessing of the optical flow vectors including removal of the rotational optical flow component, e.g. using the gyro angular rate information.

The algorithm of qualitative optical flow processing contains six steps: the *hierarchical segmentation* to obtain optical flow quadrants, the *outlier detection filter* to filter optical flow outliers, the *quadrant averaging* to calculate the averaging vector in the quadrants, the *structural identification* to reconstruct the shape of an observed object, the *fuzzy distance calculation*, and the *fuzzy-based interpretation* to extract topological environment details (topological structure from motion approach).

The proposed innovative approach for optical flow perception is performed *qualitatively* and focuses on *topological image interpretation* rather than the common execution of metric interpretation to obtain the characteristics and *structural details* of the working environment and to *recognize online* indoor landmarks.

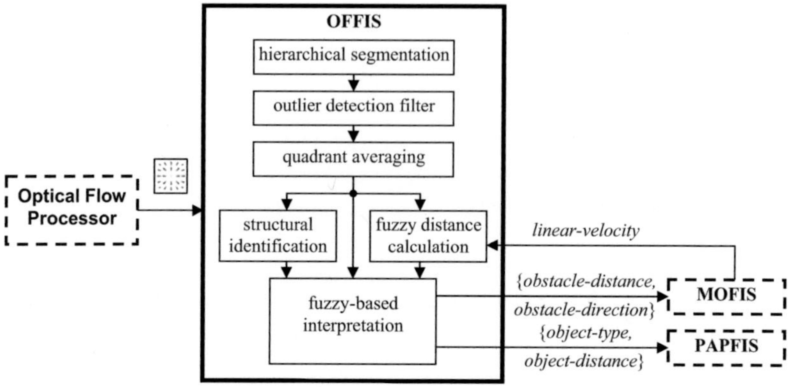

Fig. 2.2. Communication diagram of the OFFIS module

Compared to the classic metric approaches, this approach uses much simpler arithmetic and requires *less computation time* because of the use of qualitative optical flow interpretation. Moreover, averaging the optical flow patterns enables the optical flow perception to reduce to some extent the *stochastic errors* in optical flow determination and to provide for smoother 3D artifacts in the image flow.

The fuzzy-based interpretation is executed qualitatively to deliver perceptional data about observed objects (indoor landmarks), including *object-type* and *object-distance*, to the PAPFIS module and about obstacles concerning *obstacle-direction* and *obstacle-distance* to the MOFIS module.

The distance calculation is deduced from the relationship proven in (Beck, 2004) between the optical flow vector, the *linear velocity* of the robot, and the distance to the observed object.

2.3.2. PAPFIS module

The PAPFIS module executes topological path planning with two basic objectives: global path planning and local path planning. Global path planning must deal with the issues of long-term planning that are concerned with finding the best path to the target. Local path planning has to solve those problems of short-term planning that are involved in identifying local goals.

Additionally, based on the fuzzy landmark distance *object-distance* and local goal information *goal-data* of the local planner, the PAPFIS module computes primitive and composite behaviors and selects a behavior suitable for the situation at hand. The purpose of these behaviors is goal-oriented navigation that drives mobile robots approaching a local goal that is recognized through landmarks.

Fig. 2.3 illustrates the communication interfaces of the PAPFIS module with the related modules. The PAPFIS module contains three sub-modules: path planner, PAPFIS behavior, and PAPFIS behavior selector. The behaviors in the PAPFIS behavior module are organized in a hierarchical architecture.

The inputs of the PAPFIS module comprise the perceptional data, including the *object-distance* and *object-type*, the inertial data, including the *robot-direction* taken from the auxiliary sensors (Aux.), the user-defined data, including the *topological map* that represents various landmarks with a graph-based map and the *objective* that defines the start and end points, and the *linear-velocity* component of the set-point calculated by the MOFIS module.

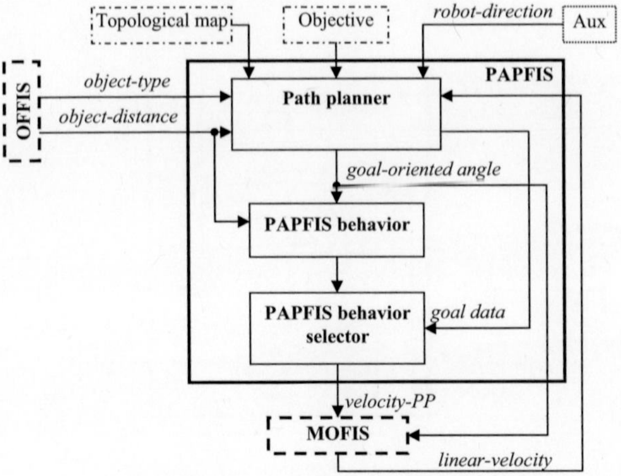

Fig. 2.3. Communication diagram of the PAPFIS module

Based on these input data, the path planner calculates a global-path from the start point to the end point of a given target, to compute the *goal-oriented angle*, and to define *goal data* for the PAPFIS behavior selector in order to activate a suitable behavior that will drive the robot to a local goal within its own range of perception.

The PAPFIS module provides the *velocity-PP* and *goal-oriented angle* for the MOFIS module.

2.3.3. MOFIS module

The MOFIS module, see Fig. 2.4, executes four main tasks: primitive behavior calculation, composite behavior calculation, MOFIS behavior selection and fuzzy-based kinematic fusion. The first three tasks are merely based on the information about the obstacles provided by the OFFIS module. The last task is not only based on local obstacle information, but also on the planning data, such as the *goal-oriented angle* and *velocity-PP* of the PAPFIS module.

The MOFIS behaviors are organized in a hierarchical architecture. The MOFIS behavior selector uses the *obstacle-distance* as decision-making information to automatically and autonomously perform a competitive coordination among the composite behaviors. The output of the selector is the velocity command *velocity-MO* for collision-free behavior.

Moreover, the MOFIS module performs a behavior fusion with the cooperative coordination of *velocity-MO* and *velocity-PP*, which are the velocity commands for collision-free behavior and goal-oriented behavior. To smooth the behavior fusion, a fuzzy inference system is applied for the cooperative coordination of the behaviors. The fuzzy inference system uses *obstacle-distance* and *goal-direction* as adjustment information to perform the fuzzy-based kinematic fusion.

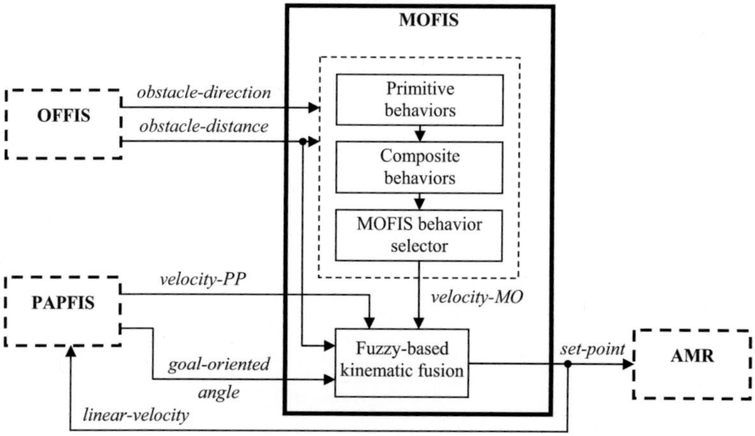

Fig. 2.4. Communication diagram of the MOFIS module

The output of the MOFIS module, which is also output of the fuzzy-based kinematic fusion, is a set-point including linear velocity and angular velocity that is provided for the locomotion of the AMR.

The three modules OFFIS, PAPFIS and MOFIS are described in detail in the next chapters.

Chapter 3
Fuzzy-based optical flow interpretation (OFFIS)

3.1. Overview

This chapter describes the new and innovative approach to bio-inspired optical flow processing in terms of a qualitative topological optical flow perception for visual mobile robots to obtain the structural information from the working environment. The optical flow perception uses fuzzy logic to make the robot perception closer to biological concepts.

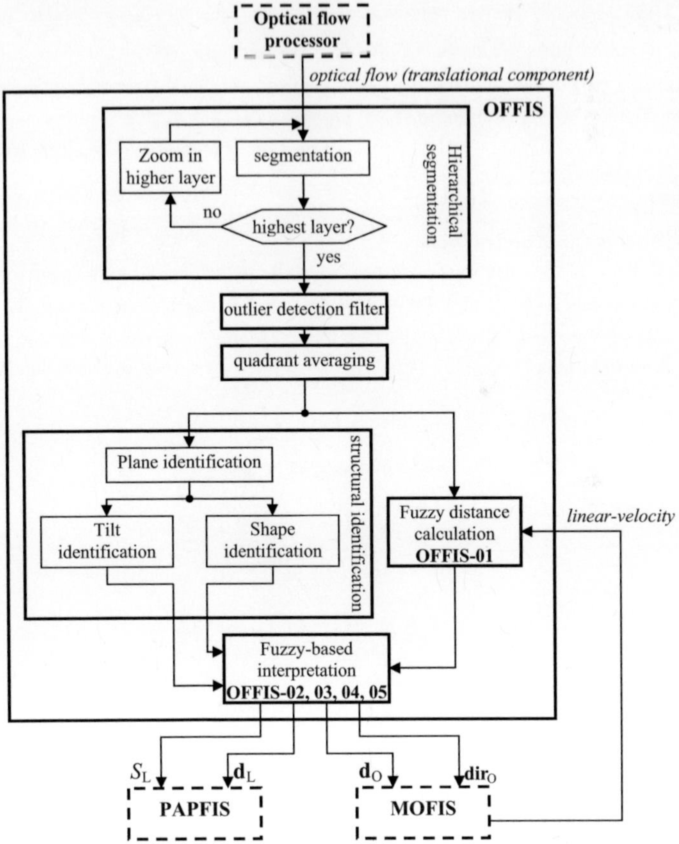

Fig. 3.1. Algorithm of optical flow perception using fuzzy logic

Compared to the classic metric approaches, this approach uses much simpler arithmetic and consumes less computation time because of the use of qualitative optical flow interpretation. The key idea of this approach is the use of *qualitative* and *topological image interpretation*, rather than the typically performed metric interpretation, for finding characteristics, to obtain *structural details* about the working environment (topological structure from motion), and to *recognize online* indoor landmarks.

The optical flow perception is executed by the OFFIS module, illustrated in Fig. 3.1, with two inputs: a translational component of the optical flow field retrieved from the optical flow processor module, and a *linear-velocity* taken from a set-point of the MOFIS module. The *linear-velocity* is only used for computing fuzzy distances.

The OFFIS module contains six sub-modules: hierarchical segmentation, outlier detection filter, quadrant averaging, structural identification, fuzzy distance calculation, and fuzzy-based interpretation. The algorithm of the optical flow perception is shown in Fig. 3.1.

The OFFIS module delivers perceptional data with four variables: obstacle distance $\mathbf{d}_{\mathrm{O}}$, obstacle direction $\mathbf{dir}_{\mathrm{O}}$, landmark distance $\mathbf{d}_{\mathrm{L}}$, and landmark type S_{L}.

In the next part of this chapter, the basic concepts of optical flow and optical flow image interpretation are briefly introduced. Then the statistical optical flow metrics are analyzed in detail. After that, further sub-module details of the OFFIS module will be given and described in the next sections.

3.2. Basic concept of optical flow

Optical flow is the distribution of the apparent velocities of movement of brightness patterns in an image (Horn and Schunck, 1981), and it arises from the relative motion of objects and an observer (camera, robot) (see Fig. 3.2). Therefore, it is possible to calculate an optical flow field from sequential image pairs captured by a mono-camera mounted on the AMR.

Fig. 3.2. Optical flow field

The observed optical flow superposes image flow from both translational and rotational camera motion. As only the translational optical flow component contains 3D information (motion stereo), it is imperative that the rotational component being removed before starting the 3D optical flow processing (Prazdny, 1980). This is most easily accomplished by using auxiliary angular rate information (e.g. gyroscope) in the same way as a fly would, compensating with its head for rotations detected by its halters (Nalbach and Hengstenberg, 1994).

For the determination of optical flow, a large number of different approaches have been investigated in the past three decades, e.g. (Beauchemin and Barron, 1995) and (Liu et al.

1998). High-performance, compact solutions for mobile robot applications have been reported recently on the basis of FPGA technology, e.g. (Diaz et al. 2006) or optical correlator technology (Tchernykh et al. 2006). Of particular interest are those methods that could provide real-time performances with some acceptable accuracy degradation for the sake of compactness and computation speed, such as a simple 1D-camera solution (Zufferey and Floreano, 2006) or a special VLSI chip (Barrows and Neely, 2000).

To cope with the bio-inspired approach, the optical flow computation should take into account an equally-spaced, but more or less sparse and corrupted optical flow field. For the investigations and experimental work discussed in the current research, a regular image segmentation with a fixed window (segment) size has been used. The determination of optical flow vectors has been performed by a 2D-correlation block-matching algorithm (Tchernykh et al. 2006).

3.3. Basic concepts of optical flow image interpretation

3.3.1. Topological structure from motion

"Structure from motion" refers to a technique to obtain the 3D structure of an object by analyzing the motion information over time based on consecutive images.

Through the measurements and interpretation of the optical flow, it is possible to infer 3D motion and position information from 2D images (Fig. 3.3).

Fig. 3.3. Structure from motion

The general optical flow classification approach follows the "structure from motion" paradigm. It uses the fact that two images captured at different locations (at two different instants assuming nonzero AMR velocity) show a spatial disparity of corresponding image points, which allows a local depth map of the observed environment to be constructed. The classical structure from motion reconstruction retrieves relative information, e.g. (Prazdny, 1980), at the expense of the considerably high computational effort and the need for additional scaling information in order to calculate a metric depth map.

The *topological structure from motion* approach proposed in this thesis uses more qualitative 3D information from the working environment and uses velocity command information to solve the scaling problem (see section 3.8.1). A fuzzy classification delivers 3D information in a topological sense, so that topographic and certain 3D object relationships, including fuzzy distance estimates, are retrieved from the optical flow patterns.

3.3.2. Hierarchical segmentation-Pyramid approach

The standard *pyramid approach* has been adopted for a *hierarchical segmentation* of the optical flow field in order to get more structural details from the surrounding environment.

The top level, layer 1, represents the four basic optical flow field quadrants (Q_1, Q_2, Q_3 and Q_4) shown in Fig. 3.4a. If more details in any layer-1 quadrant are needed, this optical flow quadrant can be further decomposed into layer-2 quadrants Q_{n1}, Q_{n2}, Q_{n3} and Q_{n4}, where n=1, 2, 3, 4 in any of the layer-1 quadrants. In Fig. 3.4b the segmentation of optical flow in layer 2 for quadrant Q_4 is sketched as an example.

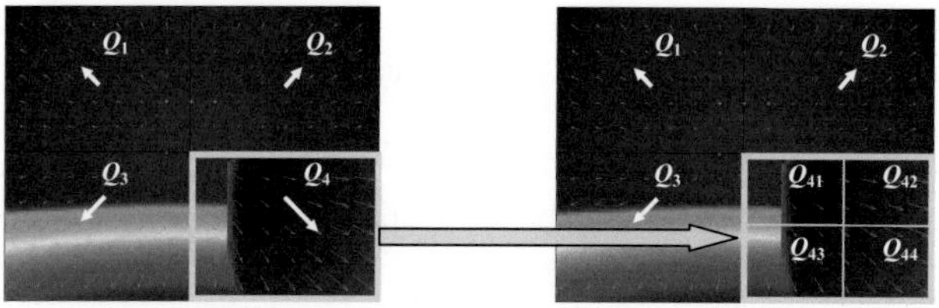

Fig. 3.4. Hierarchical segmentation of the optical flow field:
a) Segmentation in layer 1 with four layer-1 quadrants Q_1, Q_2, Q_3 and Q_4,
b) Segmentation in quadrant Q_4 with four layer-2 quadrants Q_{41}, Q_{42}, Q_{43} and Q_{44}.

In the same way, going further is enabled by zooming into layer *L*, $\{L \in \mathbb{N} \mid L \geq 3\}$. It is evident that this zooming-in feature delivers a more detailed structural insight at the cost of increased computational effort. Therefore, the value *L* should be weighted with the required computational cost and demanded accuracy of measurement.

3.3.3. Two-plane built objects (TPOs)

Walls, corners, doors and corridors are easily recognizable typical indoor objects (TIO for short) that have relatively stable shapes in most buildings. In 2D images, the TIOs are illustrated by two planes, so they are called a two-plane built object (TPO).

In the following, a two-plane built character of the TIOs will be investigated. This character is incredibly useful for robot perception as it reconstructs 3D objects from 2D images.

Assume that there are two rectangular pieces. By arranging them in a manner shown in Fig. 3.5, it is possible to recognize that these images are similar to the 2D images of a wall, a corner, a door and a corridor. For example, the simulated images of the TIOs in Fig. 3.6 are similar to the arranged images of the TPOs in Fig. 3.5. The similarity suggests an interesting solution for restructuring the TIOs from 2D images by analyzing the tilt angles of the two planes, including the left and right sides.

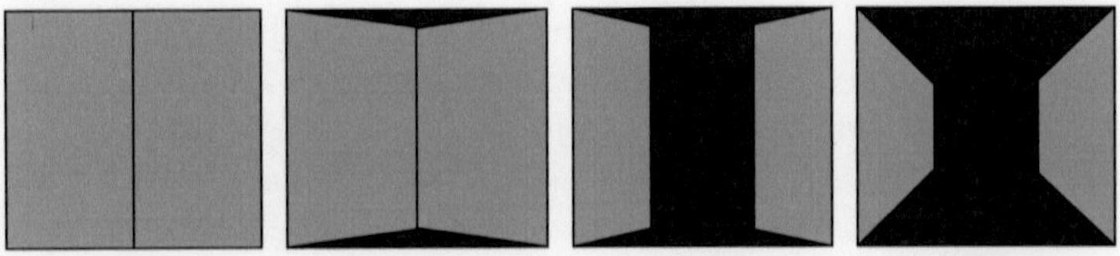

a) b) c) d)

Fig. 3.5. Arranging two planes to illustrate the TPOs: a) a wall; b) a corner; c) a door; d) a corridor

a) b) c) d)

Fig. 3.6. POV-Ray-simulated images of the TIOs: a) a wall; b) a corner; c) a door; d) a corridor

3.4. Statistical optical flow metrics

3.4.1. Averaged optical flow vector metrics

Qualitative optical flow classification is intuitively related to some averaged metrics over certain flow field regions. Biological analogies confirm this assumption and show a certain mutual interdependence between the average direction and the magnitude vs. variations of optical flow patterns. Flying insects center their flight path in corridors by balancing the left and right optical flow (Srinivasan et al. 1991), and honey bees regulate their flight speed by trying to keep the overall image motion as constant as possible (Srinivasan et al. 1996). Biomimetic experiments can now profit from the averaging of optical flow fields by gaining robustness and smoothness of motion, e.g. centering motion between average object distances (Santos-Victor et al. 1995). Averaging, in general, reduces stochastic errors in the optical flow determination and provides smoothness for 3D artifacts in the image flow.

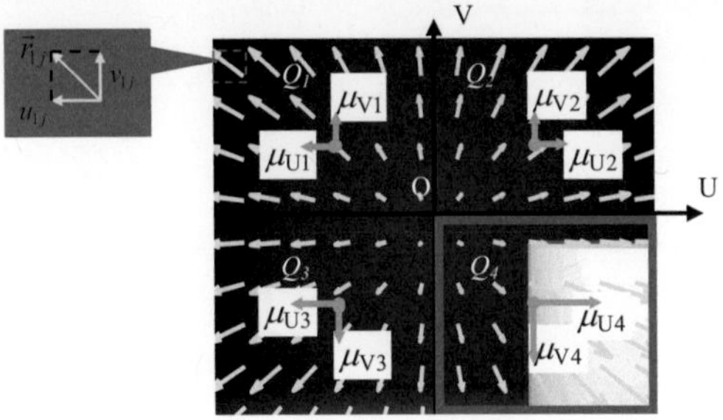

Fig. 3.7. Coordinate system of image plane and statistical values of the optical flow

Possible averaged optical flow vector metrics include classical statistical metrics like the mean (1st moment), variance (2nd moment) or optical flow divergence (Zufferey and Floreano, 2006). The latter metric was shown to be efficiently computable by using vector calculus properties in combination with special optical flow field arrangements (Poggio et al. 1991). However, it is very difficult to apply this metric to detect outliers of optical flow vectors.

For the current investigations, the classical statistical metrics, mean and standard deviation (*mean* and *stdv* for short), have been adopted for the optical flow field classification in (Mai and Janschek, 2009) as follows:

- *mean* of optical flow magnitude and direction

$$\mu_{Ui} = \frac{1}{N_i}\sum_{j=1}^{N_i}\left|u_{ij}\right|, \tag{3.1}$$

$$\mu_{Vi} = \frac{1}{N_i}\sum_{j=1}^{N_i}\left|v_{ij}\right|, \tag{3.2}$$

- *stdv* of optical flow magnitude and direction (optical flow field homogeneity)

$$\sigma_{Ui} = \sqrt{\frac{1}{N_i}\sum_{j=1}^{N_i}(\mu_{Ui}\text{-}u_{ij})^2}, \tag{3.3}$$

$$\sigma_{Vi} = \sqrt{\frac{1}{N_i}\sum_{j=1}^{N_i}(\mu_{Vi}\text{-}v_{ij})^2}, \tag{3.4}$$

where:

- i is the index of segmented quadrants, $i = 1\ldots4$,

- N_i is the number of optical flow vectors in the segmented quadrant Q_i,

- j is the index of optical flow vectors in the segmented quadrant Q_i, $j = 1\ldots N_i$,

- u_{ij} and v_{ij} are the magnitude of an optical flow vector $\vec{r}_{ij}$ in area Q_i projected respectively onto horizontal axis OU (U-axis for short) and vertical axis OV (V-axis for short) shown in Fig. 3.7),

- μ_U and μ_V are the mean of the optical flow vector magnitudes respectively on the U-axis and V-axis (*horizontal mean* and *vertical mean* for short),

- σ_U and σ_V are the *stdv* of the optical flow vector projections on the U-axis and V-axis (*horizontal stdv* and *vertical stdv* for short).

The averaging areas are chosen as the four quadrants Q_1, Q_2, Q_3 and Q_4 of the rectangular camera image, as shown in Fig. 3.7.

There are two statistical metrics available for the Pyramid approach. In the next section, the two statistical metrics in various scenes will be estimated, and then a decision will be made to select one of the two statistical metrics.

3.4.2. Statistical metrics of symmetric scenes

Fig. 3.8 displays several simple experiments to statistically compare the analytic effects of the *mean* and *stdv*. In these experiments a camera looks at the four TPOs with different distances in order to obtain *symmetric images*.

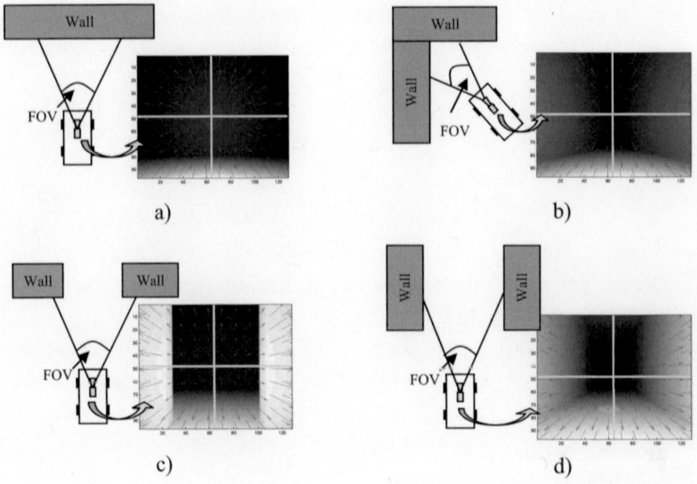

Fig. 3.8. Symmetric optical flow images of the four TPOs

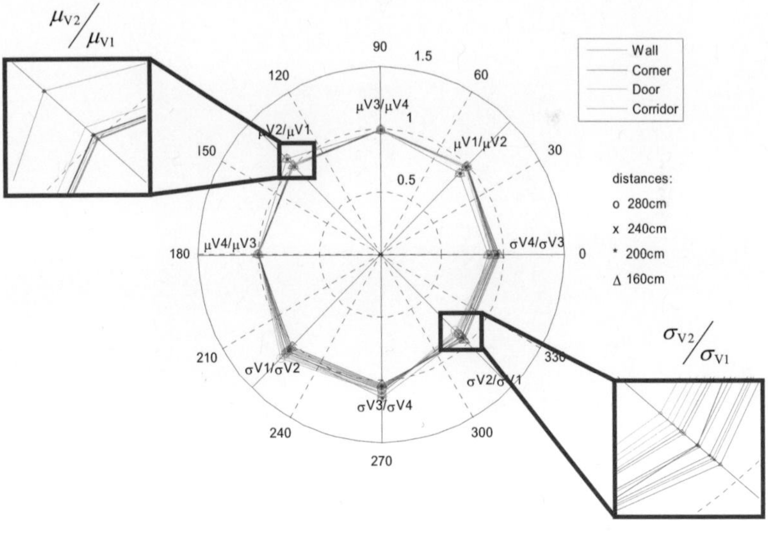

Fig. 3.9. Ratios *rmn* and *rsd* of pairs of layer-1 quadrants

The relative quantities of the *mean* and *stdv* will be considered instead of the absolute quantities of the two because the relative quantities make estimated results self-contained with various types of optical flow processing and cameras.

The ratio of the quadrant *mean* and the ratio of the quadrant *stdv* are denoted by the symbols *rmn* and *rsd*, respectively. Formally, for each pair of Q_i and Q_j, $i \neq j$, the ratios are defined

$$rmn_{ij} = \frac{\mu_{\mathrm{V}i}}{\mu_{\mathrm{V}j}}, \quad (3.5)$$

$$rsd_{ij} = \frac{\sigma_{\mathrm{V}i}}{\sigma_{\mathrm{V}j}}, \quad (3.6)$$

The results of the experiments in Fig. 3.8 are illustrated in Fig 3.9 as *rmn* and *rsd* of pairs of layer-1 quadrants. The ratios *rmn* and *rsd* in Fig. 3.9 are approximate ones because the images in Fig. 3.8 are nearly symmetric.

The above results prove that it is impossible to distinguish the TPOs based upon only layer-1 quadrants. This is obvious because the layer-1 classification only supports one plane while at least two planes are required for reconstructing the TPOs. Consequently, the following investigation of layer 2 should be taken into account.

Before investigating layer 2, two concepts, including *inside vertical mean* μ_{Vinside} and *outside vertical mean* μ_{Voutside}, should be defined.

Each layer-1 quadrant is segmented into four layer-2 quadrants, for example in Fig. 3.10, layer-1 quadrant Q_3 contains four layer-2 quadrants Q_{31}, Q_{32}, Q_{33} and Q_{34}.

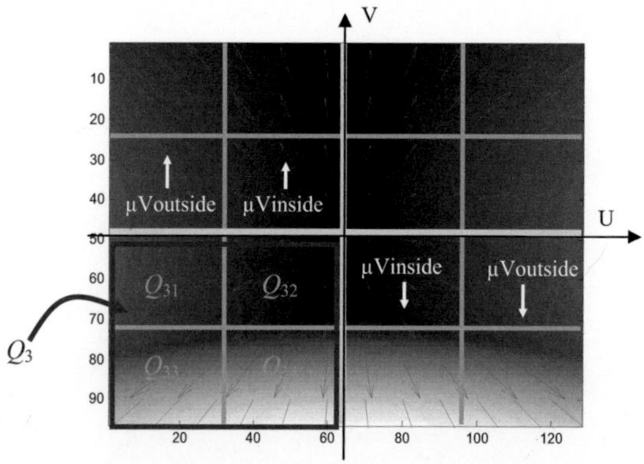

Fig. 3.10. Definition of μ_{Voutside} and μ_{Vinside}

In a comparison of relative locations to the V-axis, each layer-1 quadrant (e.g. Q_3) is divided into two columns of layer-2 quadrants, a close column (e.g. Q_{31}, Q_{33}) and a far column (e.g. Q_{32}, Q_{34}). The *mean* of the layer-2 quadrants in the close column (e.g. $\mu_{\mathrm{V}32}$, $\mu_{\mathrm{V}34}$) is referred to as the *inside vertical mean* μ_{Vinside}. Otherwise, (e.g. $\mu_{\mathrm{V}31}$, $\mu_{\mathrm{V}33}$) is the *outside vertical mean* μ_{Voutside}.

In the same way, the *inside vertical stdv* $\sigma_{\mathrm{Vinside}}$ and *outside vertical stdv* $\sigma_{\mathrm{Voutside}}$ for vertical standard deviations are defined.

By using (3.5) and (3.6), the layer-2 ratios of the *mean* and *stdv* are defined by

$$rmn = \frac{\mu_{\text{Voutside}}}{\mu_{\text{Vinside}}}, \tag{3.7}$$

$$rsd = \frac{\sigma_{\text{Voutside}}}{\sigma_{\text{Vinside}}}, \tag{3.8}$$

for each layer-1 quadrant.

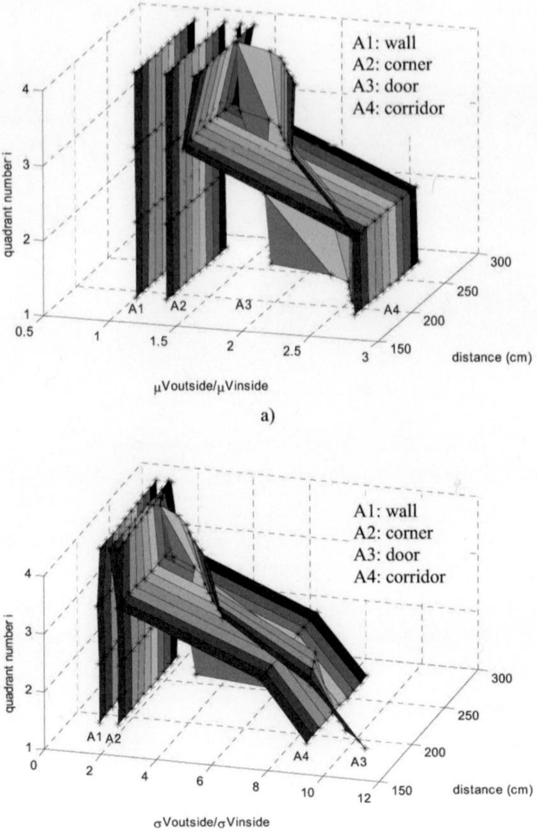

Fig. 3.11. Relationships of the layer-2 ratios of *rmn* and *rsd* with the number of layer-1 quadrants and distances in a variety of TPO situations
a) ratios of mean *rmn*, b) ratios of standard deviation *rsd*

The 3D diagram in Fig 3.11 illustrates the relationship of the layer-2 ratios of *rmn* and *rsd* in the four layer-1 quadrants with a variety of distances. These diagrams reveal the possibility of distinguishing the TPOs by the layer-2 ratios *rmn* and *rsd*. That means the

Pyramid approach requires at least two layers for the interpretations of the TPOs. Moreover, these diagrams prove that the effect of the *mean* and *stdv* are comparable in layer 2.

Actually, most optical flow images are *not symmetric*. Therefore, in the next section, the ratios of the *mean* and *stdv* in the asymmetric scenes of the TPOs will be investigated.

3.4.3. Statistical metrics of asymmetric scenes

Asymmetric images are created by placing the camera in front of the observed objects so that the line of sight does not coincide with the axis of symmetry, see Fig. 3.12 b.

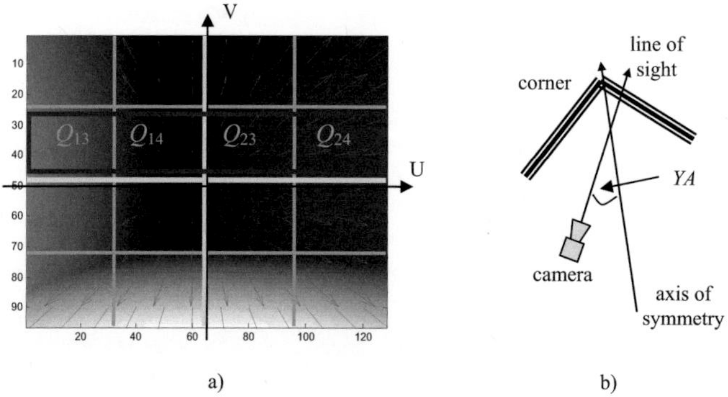

Fig. 3.12. Situation with asymmetric image

a) Optical flow with four central-row quadrants $\{Q_{13}, Q_{14}, Q_{23}, Q_{24}\}$, b) Yaw angle *YA* of camera

The following experiments will be carried out with different Yaw angles, *YA*, of the camera and various distances from the camera to the observed objects.

In contrast to the symmetric scenes, the asymmetric scenes require not one, but two layer-2 ratios rmn or rsd for TPO interpretations. For this reason, the concept of the tilt function $\mathrm{f}_{\mathrm{tilt}}$ is defined as the product of pairs of layer-2 ratios rmn or rsd. Thus,

$$\mathrm{f}_{\mathrm{tilt}}(\mu_{\mathrm{V}}) = rmn_{\mathrm{leftside}} * rmn_{\mathrm{rightside}}, \quad (3.9)$$

$$\mathrm{f}_{\mathrm{tilt}}(\sigma_{\mathrm{V}}) = rsd_{\mathrm{leftside}} * rsd_{\mathrm{rightside}}, \quad (3.10)$$

where:

- $rmn_{\mathrm{leftside}} = \dfrac{\mu_{\mathrm{Voutside\_left}}}{\mu_{\mathrm{Vinside\_left}}}$, $rmn_{\mathrm{righttside}} = \dfrac{\mu_{\mathrm{Voutside\_right}}}{\mu_{\mathrm{Vinside\_right}}}$,

- $\mu_{\mathrm{Voutside\_left}} \in \{\mu_{\mathrm{V11}}, \mu_{\mathrm{V13}}, \mu_{\mathrm{V31}}, \mu_{\mathrm{V33}}\}$, $\mu_{\mathrm{Vinside\_left}} \in \{\mu_{\mathrm{V12}}, \mu_{\mathrm{V14}}, \mu_{\mathrm{V32}}, \mu_{\mathrm{V34}}\}$,

- $\mu_{\mathrm{Voutside\_right}} \in \{\mu_{\mathrm{V22}}, \mu_{\mathrm{V24}}, \mu_{\mathrm{V42}}, \mu_{\mathrm{V44}}\}$, $\mu_{\mathrm{Vinside\_right}} \in \{\mu_{\mathrm{V21}}, \mu_{\mathrm{V23}}, \mu_{\mathrm{V41}}, \mu_{\mathrm{V43}}\}$,

- $rsd_{\mathrm{leftside}} = \dfrac{\sigma_{\mathrm{Voutside\_left}}}{\sigma_{\mathrm{Vinside\_left}}}$, $rsd_{\mathrm{righttside}} = \dfrac{\sigma_{\mathrm{Voutside\_right}}}{\sigma_{\mathrm{Vinside\_right}}}$,

- $\sigma_{\mathrm{Voutside\_left}} \in \{\sigma_{\mathrm{V11}}, \sigma_{\mathrm{V13}}, \sigma_{\mathrm{V31}}, \sigma_{\mathrm{V33}}\}$, $\sigma_{\mathrm{Vinside\_left}} \in \{\sigma_{\mathrm{V12}}, \sigma_{\mathrm{V14}}, \sigma_{\mathrm{V32}}, \sigma_{\mathrm{V34}}\}$,

- $\sigma_{\mathrm{Voutside\_right}} \in \{\sigma_{\mathrm{V22}}, \sigma_{\mathrm{V24}}, \sigma_{\mathrm{V42}}, \sigma_{\mathrm{V44}}\}$, $\sigma_{\mathrm{Vinside\_right}} \in \{\sigma_{\mathrm{V21}}, \sigma_{\mathrm{V23}}, \sigma_{\mathrm{V41}}, \sigma_{\mathrm{V43}}\}$.

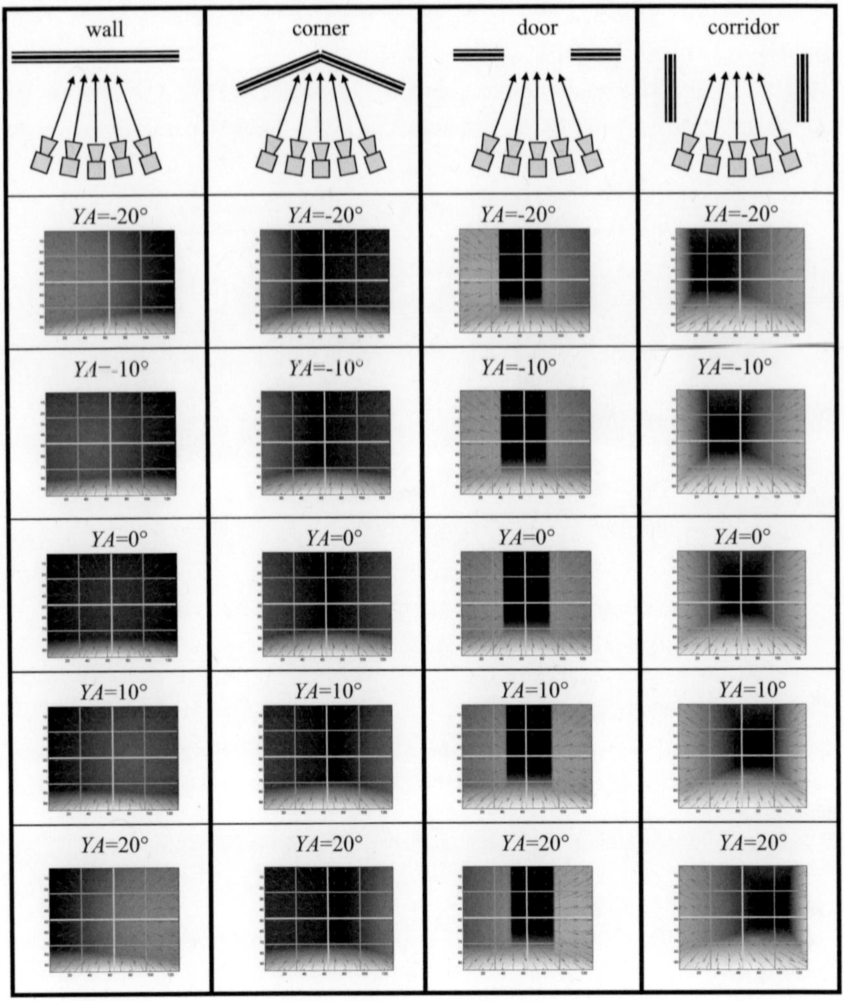

Fig. 3.13. Optical flow images of a robot encountering the four TPOs with different Yaw angles YA of camera and various distances d_{S}

Now a set of four central-row quadrants $\{Q_{13}, Q_{14}, Q_{23}, Q_{24}\}$ will be picked out as basic elements for analyzing $\mathrm{f}_{\mathrm{tilt}}(\mu_{\mathrm{V}})$ and $\mathrm{f}_{\mathrm{tilt}}(\sigma_{\mathrm{V}})$ (see Fig. 3.12 a).

To create asymmetric images, the experiments will be implemented with a set of TPO situations $S_{\mathrm{L}} = \{A_i \mid A_i \in \mathrm{TPOs}, i = 1\ldots4\}$, a set of camera Yaw angles

$YA \in \{-20°, -10°, 0°, 10°, 20°\}$, and a set of distances from the camera to the TPOs $d_S = \min(\mathbf{d}_L) = \{d_i + \Delta d_i \mid d_i \in [190, 280]cm, \Delta d_i = 10cm\}$.

Fig. 3.13 displays the optical flow images of the experiments, and Fig. 3.14 illustrates the relationships of $f_{tilt}(\mu_V)$ and $f_{tilt}(\sigma_V)$ with the Yaw angles YA and the distances d_S in the situations S_L.

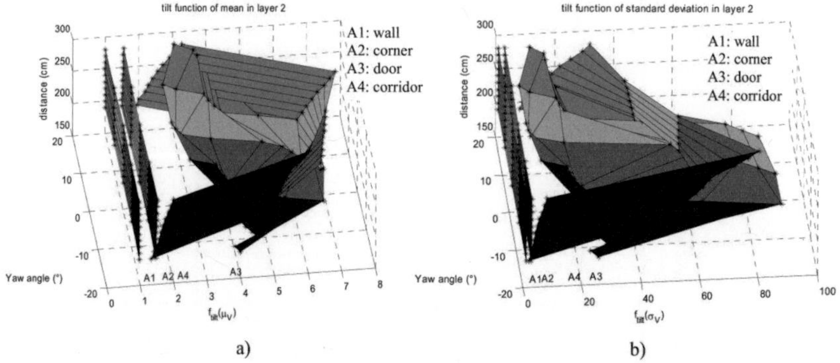

Fig. 3.14. Relationships of $f_{tilt}(\mu_V)$ and $f_{tilt}(\sigma_V)$ with different YA and d_S in variety of S_L

a) tilt function of mean $f_{tilt}(\mu_V)$, b) tilt function of standard deviation $f_{tilt}(\sigma_V)$

It is easy to recognize that $f_{tilt}(\mu_V)$ and $f_{tilt}(\sigma_V)$ are formally similar to each other and both of them can be use to distinguish the typical indoor objects. For this reason, a conclusion can be drawn that both of the statistical metrics, *mean* and *stdv*, are suitable for the Pyramid method.

The question is which of the two metrics is the preferable method for calculation?

3.4.4. Selection of statistical metrics

As the analyses described in the previous sections, the contributions of *mean* and *stdv* are similar in scale of the Pyramid approach. A demonstration will show that one of the two metrics is sufficient for the interpretation.

The redundancy of the *mean*-based classification at layer 2 and the *stdv*-based classification at layer 1 is quite evident because both classifications analyze the variations of the optical flow field in the area under consideration.

In order to weigh the pros and cons of the two metrics, the equations from (3.1) to (3.4) will be reviewed to compare the computational cost of the statistical metrics. Obviously to compute *stdv*, the equations (3.3) and (3.4) must contain the *mean* of equations (3.1) and (3.2). Therefore, the computational cost of *stdv* will always be greater than that of the *mean*.

According to the above deduction, the *mean* is preferable to *stdv*. Therefore, a multi-layer pyramidal architecture with the *mean* optical flow vector metrics has been adopted as a *baseline* for further investigations.

3.5. Hierarchical segmentation

According to the Pyramid approach, the optical flow field is segmented into layers. It is evident that the accuracy of the interpretation is directly proportional to the number of Pyramid layers. The computational cost is, however, also directly proportional to the number of Pyramid layers. Through the experimental investigations of the current work, it is recognized that a Pyramid with three layers will yield reliable classification results. Therefore, the hierarchical segmentation with three layers is selected in this research.

After the first segmentation in layer 1, the optical flow field is divided into four quadrants. These quadrants are arranged in a (2x2) matrix as the following:

$$\mathbf{\Omega}^{\mathrm{I}} = \begin{bmatrix} Q_1 & Q_2 \\ Q_3 & Q_4 \end{bmatrix}. \tag{3.11}$$

Next, the four layer-1 quadrants are divided into layer-2 quadrants as shown in Fig. 3.15. That means there are 16 layer-2 quadrants in layer 2. The (4x4) matrix of the layer-2 quadrants is formed as

$$\mathbf{\Omega}^{\mathrm{II}} = \begin{bmatrix} Q_{11} & Q_{12} & Q_{21} & Q_{22} \\ Q_{13} & Q_{14} & Q_{23} & Q_{24} \\ Q_{31} & Q_{32} & Q_{41} & Q_{42} \\ Q_{33} & Q_{34} & Q_{43} & Q_{44} \end{bmatrix}. \tag{3.12}$$

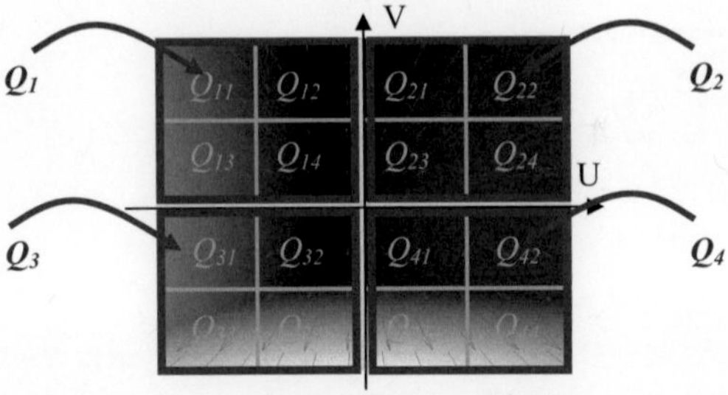

Fig. 3.15. Layer-2 quadrants segmented from four layer-1 quadrants

Similarly, the (8x8) matrix of the layer-3 quadrants is

$$\mathbf{\Omega}^{\mathrm{III}} = \begin{bmatrix} Q_{111} & Q_{112} & Q_{121} & Q_{122} & Q_{211} & Q_{212} & Q_{221} & Q_{222} \\ Q_{113} & Q_{114} & Q_{123} & Q_{124} & Q_{213} & Q_{214} & Q_{223} & Q_{224} \\ Q_{131} & Q_{132} & Q_{141} & Q_{142} & Q_{231} & Q_{232} & Q_{241} & Q_{242} \\ Q_{133} & Q_{134} & Q_{143} & Q_{144} & Q_{233} & Q_{234} & Q_{243} & Q_{244} \\ Q_{311} & Q_{312} & Q_{321} & Q_{322} & Q_{411} & Q_{412} & Q_{421} & Q_{422} \\ Q_{313} & Q_{314} & Q_{323} & Q_{324} & Q_{413} & Q_{414} & Q_{423} & Q_{424} \\ Q_{331} & Q_{332} & Q_{341} & Q_{342} & Q_{431} & Q_{432} & Q_{441} & Q_{442} \\ Q_{333} & Q_{334} & Q_{343} & Q_{344} & Q_{433} & Q_{434} & Q_{443} & Q_{444} \end{bmatrix} \tag{3.13}$$

Each quadrant Q_i contains two *mean* metrics, including $\mu_{\mathrm{U}i}$ and $\mu_{\mathrm{V}i}$, specifically described in section 3.4.

3.6. Quadrant averaging

In any quadrant, the average of the optical flow vectors mapping on the U-axis is calculated by equation (3.1), and mapping on the V-axis is calculated by equation (3.2).

3.7. Structural identification

Based on the results of the hierarchical segmentation, a structural identification is executed by three tasks: plane identification answers the question, "Is there a pair of right and left planes?" Tilt calculation responds to the question, "How big are the tilt angles of the planes?" Finally, shape interpretation answers the question, "What does the optical flow image look like?" More details about the three tasks will be described in the subsequent sections.

3.7.1. Plane identification

Reconsider Fig. 3.15, it is obvious that if Q_{11} and Q_{13} cooperatively contain a vertical plane, they certainly have the same average values of *horizontal mean*, in other words μ_{U11} is equal to μ_{U13}. In a similar manner, if Q_{22} and Q_{24} collectively hold a vertical plane, μ_{U22} is equal to μ_{U24}.

The vertical plane function is defined by the following formula:

$$\text{vplane}(Q_i, Q_j) = \begin{cases} 0 \Leftrightarrow \mu_{Ui} \neq \mu_{Uj} \\ 1 \Leftrightarrow \mu_{Ui} = \mu_{Uj} \end{cases}, \qquad (3.14)$$

and the horizontal plane function is

$$\text{hplane}(Q_i, Q_j) = \begin{cases} 0 \Leftrightarrow \mu_{Vi} \neq \mu_{Vj} \\ 1 \Leftrightarrow \mu_{Vi} = \mu_{Vj} \end{cases}. \qquad (3.15)$$

Generally the plane function can be expressed as

$$\text{plane}(Q_i, Q_j) = \begin{cases} 0 \Leftrightarrow \mu_i \neq \mu_j \\ 1 \Leftrightarrow \mu_i = \mu_j \end{cases}. \qquad (3.16)$$

The experimental investigations have shown that equation (3.16) is *not satisfactory* in real environments because of surface relief or inhomogeneous structure (e.g. Q_3 and Q_4 in Fig. 3.15 contain a floor) and/or the noise of optical flow processing.

To deal with this problem, equation (3.16) has been modified as

$$\text{plane}(Q_i, Q_j) = \begin{cases} 0 & \text{if } \left(\mu_i / \mu_j < 1 / k_p \right) \text{ OR } \left(\mu_i / \mu_j > k_p \right) \\ 1 & \text{otherwise} \end{cases}, \qquad (3.17)$$

where k_p is the threshold of the plane defined through experiments, e.g. $k_p = 1.2$.

3.7.2. Tilt identification

First of all, the tilt angle concept should be clarified. Tilt angle is an angle (φ in Fig. 3.16) between the *line of sight* plane LOS and the *flat-object* plane FO, illustrated in Fig. 3.16. The plane LOS is a vertical plane ($\angle$ZOY) on the direct line between an observer (camera, robot) and a target (object). The plane FO is a vertical plane ($\angle$ZOX) that contains a flat object like a wall.

Similar to the ratio calculated by equation (3.7), the tilt of a plane is defined as

$$\mathrm{tilt}(Q_i, Q_j) = \frac{\mu_{Vi}}{\mu_{Vj}}, \tag{3.18}$$

where μ_{Vi} belongs to the *outside vertical mean* and μ_{Vj} belongs to the *inside vertical mean*, (see again Fig. 3.10).

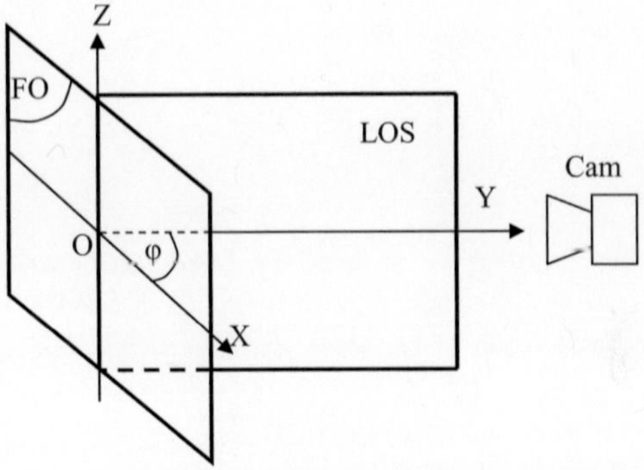

Fig. 3.16. Definition of tilt angle

3.7.3. Shape interpretation

Based on the matrixes of layer-2 quadrants $\mathbf{\Omega}^{\mathrm{II}}$ and layer-3 quadrants $\mathbf{\Omega}^{\mathrm{III}}$, shape interpretation is divided into rough shape interpretation and fine interpretation.

Fig. 3.17. Layer-2 quadrants for rough shape interpretation

Rough shape interpretation executes on layer-2 quadrants. By comparing the layer-2 quadrants of a layer-1 quadrant from each other, the *closest quadrant* with the biggest average value is able to be found. Subsequently, using the results of plane identification, the *closest plane* containing the closest quadrant can be identified. The closest planes form a rough shape with a matrix of 1 and 0.

The situation illustrated in Fig. 3.17 is explained as an example of rough shape interpretation. Firstly, by comparing layer-2 quadrants in layer-1 quadrant Q_1, it is possible to figure out Q_{11} is the closest quadrant in Q_1. If plane$(Q_{11}, Q_{13}) = 1$, plane(Q_{11}, Q_{13}) is the closest area in Q_1 because it covers the closest quadrant Q_{11}.

In the same way, the closest areas of layer 2 are plane(Q_{22}, Q_{24}) in Q_2, plane(Q_{31}, Q_{33}) and plane(Q_{33}, Q_{34}) in Q_3 , and plane(Q_{42}, Q_{44}) and plane(Q_{43}, Q_{44}) in Q_4.

Symbolizing the closest areas with 1, and others with 0, the matrix of rough shape of the considered situation can be formed as

$$\mathbf{roughShape} = \begin{bmatrix} 1 & 0 & 0 & 1 \\ 1 & 0 & 0 & 1 \\ 1 & 0 & 0 & 1 \\ 1 & 0 & 0 & 1 \end{bmatrix}. \tag{3.19}$$

Fine shape interpretation is identified by layer-3 quadrants. Similar to rough shape interpretation, a fine shape is constructed with a matrix of 1 and 0.

Upon the reconsideration of the above example shown in Fig. 3.17 and the use of the matrix of layer-3 quadrants $\mathbf{\Omega}^{\mathrm{III}}$, the matrix of fine shape is

$$\mathbf{fineShape} = \begin{bmatrix} 1 & 1 & 0 & 0 & 0 & 0 & 0 & 1 \\ 1 & 1 & 0 & 0 & 0 & 0 & 0 & 1 \\ 1 & 1 & 0 & 0 & 0 & 0 & 0 & 1 \\ 1 & 1 & 0 & 0 & 0 & 0 & 0 & 1 \\ 1 & 1 & 0 & 0 & 0 & 0 & 0 & 1 \\ 1 & 1 & 0 & 0 & 0 & 0 & 0 & 1 \\ 1 & 1 & 0 & 0 & 0 & 0 & 0 & 1 \\ 1 & 1 & 0 & 0 & 0 & 0 & 0 & 1 \end{bmatrix}. \tag{3.20}$$

The rough and fine shape interpretations support structural details of the working environment considered to be topological structure from motion.

3.8. Fuzzy distance calculation

As mentioned in section 3.2, the optical flow field contains translational and rotational components related to the translational and rotational motion of the camera. Actually, 3D information (motion stereo) can only be retrieved from translational optical flow components (Prazdny, 1980). The removal of rotational optical flow components is described, e.g. in (Nalbach and Hengstenberg, 1994).

In the following considerations, the geometric relationship between the translational optical flow component and the distance from a camera to an observed object will be depicted. Then a fuzzy distance calculation will be described.

3.8.1. Principle of optical flow-based distance calculation

Illustrated in Fig. 3.18, the geometric relationship between the translational optical flow component and the distance to an observed object is taken into account in two coordinate systems: XYZ for camera and UVZ for image plane. The image plane is defined as a virtual plane on which an image is projected. The Z-axis of the camera coordinate system concurs with the optical axis of the camera. Additionally, the image plane is assumed to be in front of the camera with the distance f that is the *focal length* of the camera.

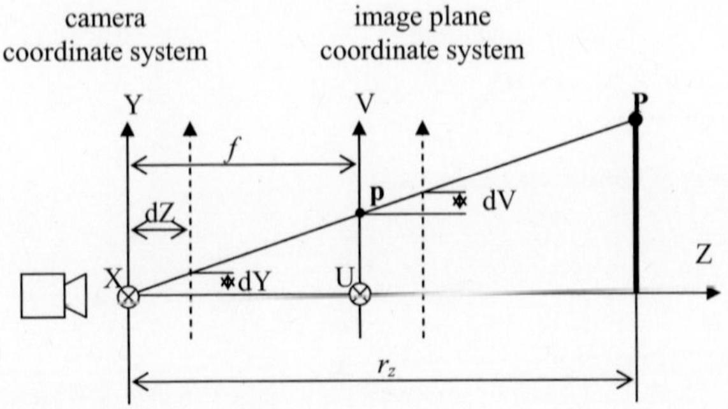

Fig. 3.18. Geometric relationship between the translational optical flow component and the distance to observed object **P** on the Z-axis of the camera coordinate system (pinhole projection)

Assume that there is an arbitrary object in front of the camera with a distance r_z. The top point of the observed object is at $\mathbf{P} = [X_P \quad Y_P \quad Z_P]^T$. This point is projected onto the image plane by point $\mathbf{p} = [U \quad V \quad Z]^T$. The position of **p** is represented

$$\frac{\mathbf{p}}{Z} = \frac{\mathbf{P}}{Z_P} \Leftrightarrow \frac{\mathbf{p}}{f} = \frac{\mathbf{P}}{r_z}. \tag{3.21}$$

The distance r_z changes after camera movement, so the temporal differential of **p** is calculated

$$\dot{\mathbf{p}} = \frac{\dot{\mathbf{P}}}{r_z} \cdot f - \mathbf{P} \cdot \frac{\dot{r}_z}{r_z^2} \cdot f. \tag{3.22}$$

Suppose that the camera moves with the translational velocity $\mathbf{v} = [v_x \quad v_y \quad v_z]^T$ and the angular velocity $\boldsymbol{\omega} = [\omega_x \quad \omega_y \quad \omega_z]^T$ so that the velocity of the point **P** is identified by

$$\dot{\mathbf{P}} = -\boldsymbol{\omega} \times \mathbf{P} - \mathbf{v}. \tag{3.23}$$

From equations (3.21), (3.22) and (3.23), the optical flow is calculated by

$$\dot{\mathbf{p}} = -\left(\boldsymbol{\omega} \times \mathbf{p} + \mathbf{v} + \mathbf{p} \cdot \frac{\dot{r}_z}{r_z} \right). \tag{3.24}$$

Equation (3.24) illustrates the velocity of an image point, which is called optical flow vector. Because the image plane has two dimensions, henceforward any optical flow vector is to be considered a vector of two elements projected on the U- and V-axis $\dot{\mathbf{p}} = [u \quad \upsilon]^T$, respectively.

There are many publications, e.g. (Beck, 2004) and (Shibata et al. 2008), which describe solutions for equation (3.24). The solution in (Shibata et al. 2008) has been selected for further calculation. According to this solution, the optical flow vector is represented as

$$\begin{bmatrix} u \\ \upsilon \end{bmatrix} = \mathbf{M}_{\text{Rot}} \cdot \begin{bmatrix} \omega_x \\ \omega_y \\ \omega_z \end{bmatrix} + \mathbf{M}_{\text{Trans}} \cdot \begin{bmatrix} v_x \\ v_y \\ v_z \end{bmatrix}, \tag{3.25}$$

where:

- $\mathbf{M}_{\text{Rot}}$ is a matrix of rotational components

$$\mathbf{M}_{\text{Rot}} = \begin{bmatrix} U \cdot V & -\left(U^2+1\right) & -V \\ V^2+1 & -U \cdot V & U \end{bmatrix},$$

- $\mathbf{M}_{\text{Trans}}$ is a matrix of translational components

$$\mathbf{M}_{\text{Trans}} = \begin{bmatrix} -\frac{1}{r_z} & 0 & \frac{U}{r_z} \\ 0 & -\frac{1}{r_z} & \frac{V}{r_z} \end{bmatrix}.$$

Rotational components can be easily removed if the angular velocity ω is known by measurements (e.g. gyro) or by the commands to the motion control. Therefore,

$$\begin{bmatrix} \hat{u} \\ \hat{\upsilon} \end{bmatrix} = \mathbf{M}_{\text{Trans}} \cdot \begin{bmatrix} v_x \\ v_y \\ v_z \end{bmatrix} = \begin{bmatrix} \tilde{u} \\ \tilde{\upsilon} \end{bmatrix} - \mathbf{M}_{\text{Rot}} \cdot \begin{bmatrix} \omega_x \\ \omega_y \\ \omega_z \end{bmatrix}. \tag{3.26}$$

where

- $\hat{u}$ and $\hat{\upsilon}$ are respectively estimated values of optical flow on the U- and V-axis,
- $\tilde{u}$ and $\tilde{\upsilon}$ are respectively measured values of optical flow on the U- and V-axis.

Equation (3.26) can be simply rewritten as

$$\hat{u} = -\frac{1}{r_z} \cdot v_x + \frac{U}{r_z} \cdot v_z, \tag{3.27}$$

$$\hat{\upsilon} = -\frac{1}{r_z} \cdot v_y + \frac{V}{r_z} \cdot v_z. \tag{3.28}$$

With the assumption that a ground-based mobile robot mainly moves forward without swinging motions, e.g. jumping up/down or slipping across, the velocities would be v_x and $v_y \ll v_z$. Stemming from equations (3.27) and (3.28), the distance r_z can be calculated

$$r_z = \frac{U}{\hat{u}} \cdot v_z, \tag{3.29}$$

$$r_z = \frac{V}{\hat{\upsilon}} \cdot v_z. \tag{3.30}$$

Equations (3.29) and (3.30) support two ways to calculate distance r_z from optical flow vectors. It is possible to select one of them by comparing v_x and v_y. For example, if a ground-based mobile robot moves with velocities $v_x \gg 0$ and $v_y \approx 0$, equation (3.28) should be selected because one element is zero, i.e. equation (3.30) will be selected.

Actually, to reduce stochastic errors, a feasible solution is proposed using the average value of optical flow vectors in an optical flow pattern instead of the distance calculation in each vector, i.e. $\hat{u}$ and $\hat{\upsilon}$ are substituted by the mean metrics μ_U and μ_V, and using an estimated velocity instead of the real linear velocity v_z. Due to the substitutions, the distance r_z can be calculated by

$$r_z = \frac{U_\mu}{\mu_U} \cdot \hat{v}_z = \frac{V_\mu}{\mu_V} \cdot \hat{v}_z = \frac{1}{2} \cdot \left(\frac{U_\mu}{\mu_U} + \frac{V_\mu}{\mu_V} \right) \cdot \hat{v}_z, \tag{3.31}$$

where:

- $\hat{v}_z$ is the linear velocity estimated from the velocity commands to motion control,
- μ_U and μ_V are respectively the *mean* of optical flow on the U-axis and V-axis,
- U_μ and V_μ are respectively the coordinates of the central point of the averaged optical flow pattern.

Equation (3.31) makes it possible to calculate the object distance from the optical flow vectors. Unfortunately, a considerable level of noise is created as a result of the imaging process (e.g. lightning conditions, image sensor noise) and/or properties of the working environment. Because of this, the optical flow-based measurement of object distance is rather inaccurate. This can be taken into account by using a fuzzy logic approach (design of membership functions). Further details concerning fuzzy distance calculation are described in the next section.

3.8.2. Fuzzy-based distance calculation

The module for fuzzy distance calculation is illustrated in Fig. 3.19. The inputs of this module are the *mean* metrics μ_U and μ_V, the coordinates of the central point of an optical flow quadrant U_μ and V_μ, and the estimated linear velocity $\hat{v}_z$.

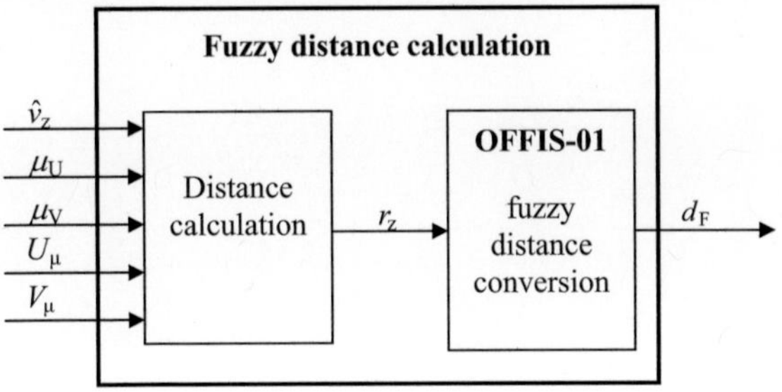

Fig. 3.19. Module of fuzzy distance calculation

The sub-module of distance calculation performs equation (3.31) to calculate crisp the distance $r_z \in \mathbb{R}^+$. The sub-module of fuzzy distance conversion OFFIS-01 transforms crisp distance r_z to fuzzy distance d_F, $0 \le d_F \le 1$ by fuzzy rules listed in Fig. 3.20. The membership function of the input r_z and output d_F are respectively illustrated in Fig. 3.21 and Fig. 3.22.

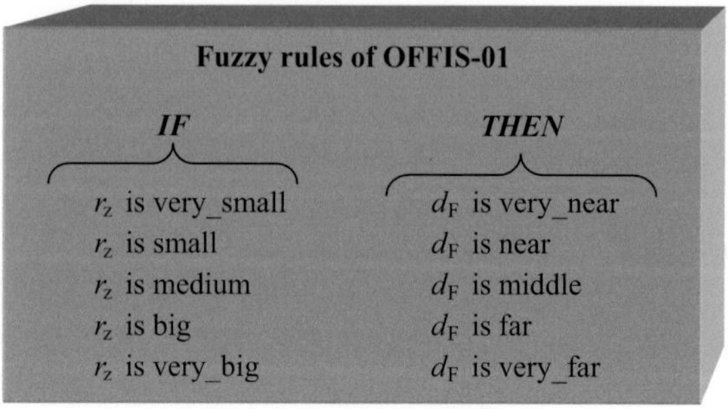

Fig. 3.20. Fuzzy rules of OFFIS-01

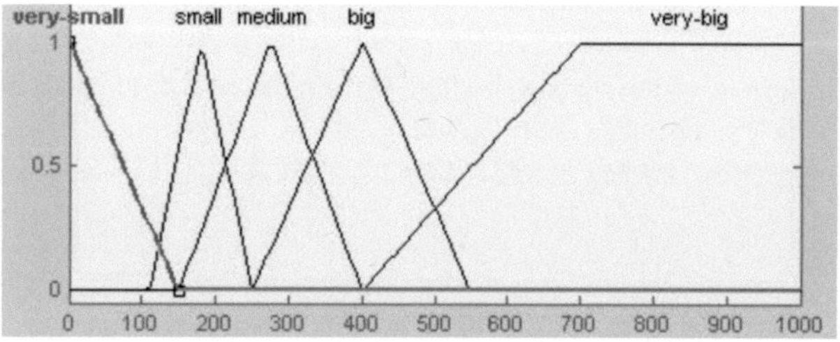

Fig. 3.21. Membership function of the input r_z of OFFIS-01

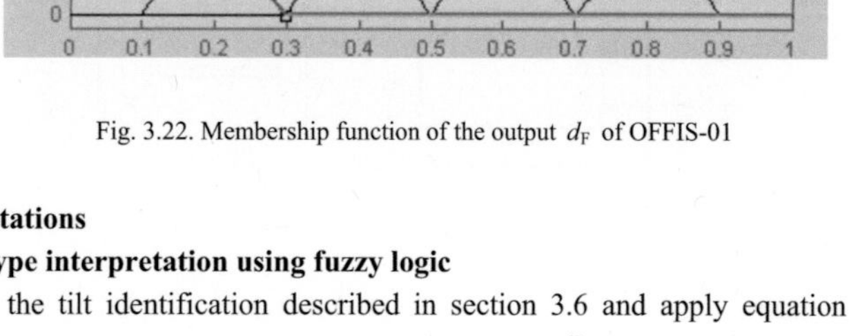

Fig. 3.22. Membership function of the output d_F of OFFIS-01

3.9. Interpretations

3.9.1. TPO type interpretation using fuzzy logic

Reconsider the tilt identification described in section 3.6 and apply equation (3.18) to compute the tilt of the layer-2 quadrants in the matrix $\mathbf{\Omega}^{\mathrm{II}}$, the matrix of layer-2 tilts is defined

$$\mathbf{tiltL2} = \begin{bmatrix} \mathrm{tilt}(Q_{11}, Q_{12}) & \mathrm{tilt}(Q_{22}, Q_{21}) \\ \mathrm{tilt}(Q_{13}, Q_{14}) & \mathrm{tilt}(Q_{24}, Q_{23}) \\ \mathrm{tilt}(Q_{31}, Q_{32}) & \mathrm{tilt}(Q_{42}, Q_{41}) \\ \mathrm{tilt}(Q_{33}, Q_{34}) & \mathrm{tilt}(Q_{44}, Q_{43}) \end{bmatrix}. \tag{3.32}$$

Also, when applying equation (3.18) for layer-3 quadrants and using only the two first and two last columns of the matrix the matrix $\mathbf{\Omega}^{\mathrm{III}}$, the matrix of layer-3 tilts is

$$\mathbf{tiltL3}=\begin{bmatrix} \mathrm{tilt}(Q_{111},Q_{112}) & \mathrm{tilt}(Q_{222},Q_{221}) \\ \mathrm{tilt}(Q_{113},Q_{114}) & \mathrm{tilt}(Q_{224},Q_{223}) \\ \mathrm{tilt}(Q_{131},Q_{132}) & \mathrm{tilt}(Q_{242},Q_{241}) \\ \mathrm{tilt}(Q_{133},Q_{134}) & \mathrm{tilt}(Q_{244},Q_{243}) \\ \mathrm{tilt}(Q_{311},Q_{312}) & \mathrm{tilt}(Q_{422},Q_{421}) \\ \mathrm{tilt}(Q_{313},Q_{314}) & \mathrm{tilt}(Q_{424},Q_{423}) \\ \mathrm{tilt}(Q_{331},Q_{332}) & \mathrm{tilt}(Q_{442},Q_{441}) \\ \mathrm{tilt}(Q_{333},Q_{334}) & \mathrm{tilt}(Q_{444},Q_{443}) \end{bmatrix}. \quad (3.33)$$

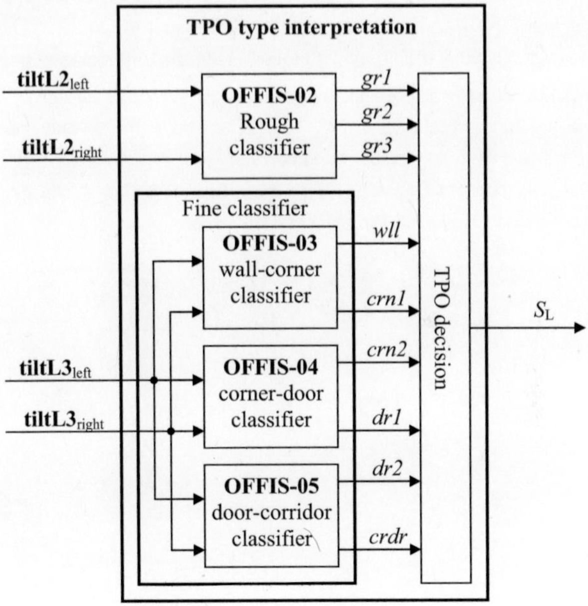

Fig. 3.23. Module of TPO type interpretation using fuzzy inference systems

When interpreting TPO type, not all pairs of tilts in **tiltL2** and **tiltL3** are necessary. Using the results of shape interpretation, described in section 3.6, including the **roughShape** and **fineShape**, keep available tilts as the following:

$$\mathbf{tiltL2}_{\mathrm{left}} = \mathbf{tiltL2}(:,1) * \mathbf{roughShape}(:,1), \quad (3.34)$$

$$\mathbf{tiltL2}_{\mathrm{right}} = \mathbf{tiltL2}(:,end) * \mathbf{roughShape}(:,end), \quad (3.35)$$

$$\mathbf{tiltL3}_{\mathrm{left}} = \mathbf{tiltL3}(:,1) * \mathbf{fineShape}(:,1), \quad (3.36)$$

$$\mathbf{tiltL3}_{\mathrm{right}} = \mathbf{tiltL3}(:,end) * \mathbf{fineShape}(:,end), \quad (3.37)$$

where A*B denotes the multiplication of each scalar component of A by the corresponding magnitude of the component vector of B.

The module of TPO type interpretation is illustrated in Fig. 3.23 with four fuzzy inference systems, including OFFIS-02, 03, 04, and 05.

The OFFIS-02 module plays the role of a rough classifier with two inputs, $\mathbf{tiltL2}_{\text{left}}$ and $\mathbf{tiltL2}_{\text{right}}$, for classifying a TPO into the three groups *gr1*, *gr2*, and *gr3*. Where, *gr1*, *gr2* and *gr3* are the probabilities of TPO $\in$ {wall, corner}, {corner, door}, and {door, corridor}, respectively.

The membership functions of the inputs and outputs of OFFIS-02 are respectively illustrated in Fig. 3.24 and Fig. 3.25. The IF-THEN fuzzy rules of OFFIS-02 are displayed in Fig. 3.26 with $tL2\_left \in \mathbf{tiltL2}_{\text{left}}$ and $tL2\_right \in \mathbf{tiltL2}_{\text{right}}$.

The fine classifier module comprises the three fuzzy inference systems OFFIS-03, OFFIS-04, and OFFIS-05 to distinguish a landmark in a pair of the TPOs. The inputs of these classifiers are $\mathbf{tiltL3}_{\text{left}}$ and $\mathbf{tiltL3}_{\text{right}}$ with membership functions shown in Fig. 3.27.

The fine classifier has six outputs: *wll*, *crn1*, *crn2*, *dr1*, *dr2* and *crdr*. The membership function of these outputs is shown in Fig. 3.28. Where, *wll*, *crn*, *dr* and *crdr* are respectively the potentialities of a wall, corner, door and corridor.

Notice that *crn1* and *crn2* are also the potentialities of a corner, and *dr1* and *dr2* are the potentialities of a door. However, *crn1* $\neq$ *crn2*, and *dr1* $\neq$ *dr2*.

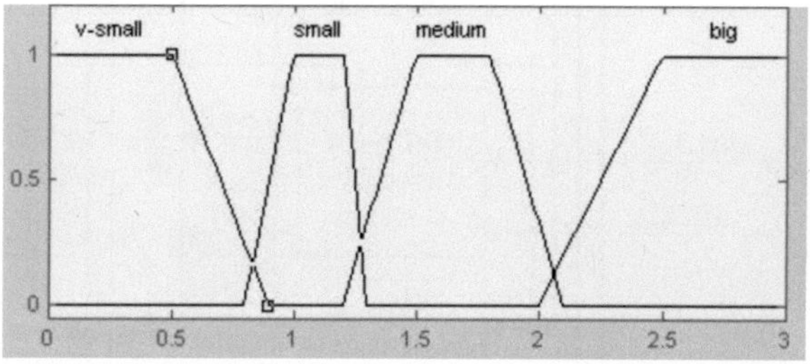

Fig. 3.24. Membership function of inputs of OFFIS-02

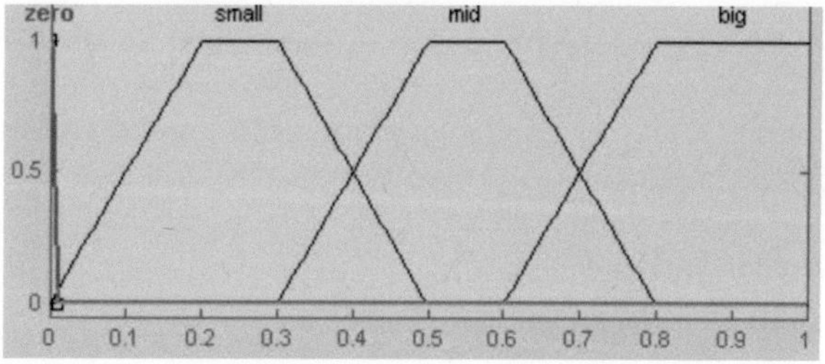

Fig. 3.25. Membership function of outputs of OFFIS-02

Fuzzy rules of OFFIS-02

IF	*THEN*
tL2_left v-small OR *tL2_right* v-small	*gr1* big AND *gr2* zero AND *gr3* zero
tL2_left small AND *tL2_right* small	*gr1* big AND *gr2* small AND *gr3* zero
tL2_left small AND *tL2_right* medium	*gr1* small AND *gr2* big AND *gr3* zero
tL2_left small AND *tL2_right* big	*gr1* zero AND *gr2* big AND *gr3* small
tL2_left medium AND *tL2_right* small	*gr1* small AND *gr2* big AND *gr3* zero
tL2_left medium AND *tL2_right* medium	*gr1* zero AND *gr2* big AND *gr3* mid
tL2_left medium AND *tL2_right* big	*gr1* zero AND *gr2* small AND *gr3* big
tL2_left big AND *tL2_right* small	*gr1* zero AND *gr2* big AND *gr3* small
tL2_left big AND *tL2_right* medium	*gr1* zero AND *gr2* small AND *gr3* big
tL2_left big AND *tL2_right* big	*gr1* zero AND *gr2* zero AND *gr3* big

Fig. 3.26. Fuzzy rules of OFFIS-02

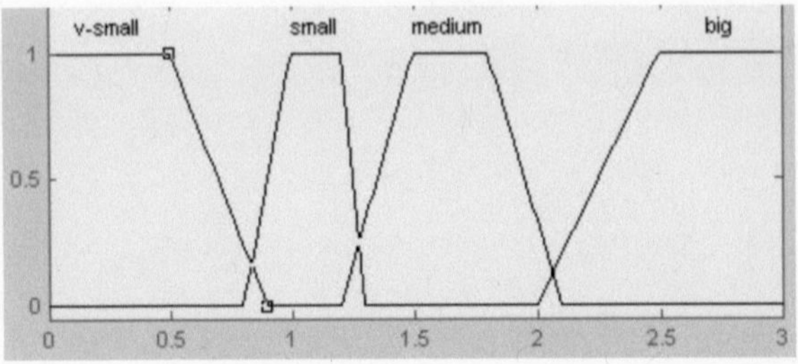

Fig. 3.27. Membership function of inputs of OFFIS-03, 04, and 05

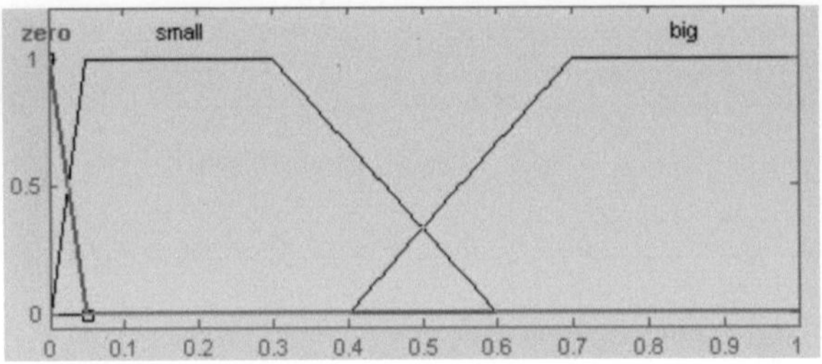

Fig. 3.28. Membership function of outputs of OFFIS-03, 04, and 05

The OFFIS-03 module is responsible for distinguishing walls and corners. The fuzzy rules of the OFFIS-03 are shown in Fig. 3.29 with $tL3\_left \in \mathbf{tiltL3}_{\mathrm{left}}$ and $tL3\_right \in \mathbf{tiltL3}_{\mathrm{right}}$.

Similarly, the OFFIS-04 module is responsible for distinguishing corners and doors and the OFFIS-05 module is responsible for distinguishing doors and corridors. The fuzzy rules of the OFFIS-04 and OFFIS-05 are shown in Fig. 3.30 and Fig. 3.31, respectively.

Fuzzy rules of OFFIS-03

IF	*THEN*
tL3_left v-small OR *tL3_right* v-small	*wll* is big AND *crn1* is zero
tL3_left small AND *tL3_right* small	*wll* is big AND *crn1* is small
tL3_left small AND *tL3_right* medium	*wll* is small AND *crn1* is big
tL3_left small AND *tL3_right* big	*wll* is small AND *crn1* is big
tL3_left medium AND *tL3_right* small	*wll* is small AND *crn1* is big
tL3_left medium AND *tL3_right* medium	*wll* is zero AND *crn1* is big
tL3_left medium AND *tL3_right* big	*wll* is zero AND *crn1* is zero
tL3_left big AND *tL3_right* small	*wll* is small AND *crn1* is big
tL3_left big AND *tL3_right* medium	*wll* is zero AND *crn1* is zero
tL3_left big AND *tL3_right* big	*wll* is zero AND *crn1* is zero

Fig. 3.29. Fuzzy rules of OFFIS-03

Fuzzy rules of OFFIS-04

IF	*THEN*
tL3_left v-small OR *tL3_right* v-small	*crn2* is zero AND *dr1* is zero
tL3_left small AND *tL3_right* small	*crn2* is big AND *dr1* is small
tL3_left small AND *tL3_right* medium	*crn2* is big AND *dr1* is small
tL3_left small AND *tL3_right* big	*crn2* is small AND *dr1* is big
tL3_left medium AND *tL3_right* small	*crn2* is big AND *dr1* is small
tL3_left medium AND *tL3_right* medium	*crn2* is small AND *dr1* is big
tL3_left medium AND *tL3_right* big	*crn2* is zero AND *dr1* is big
tL3_left big AND *tL3_right* small	*crn2* is small AND *dr1* is big
tL3_left big AND *tL3_right* medium	*crn2* is zero AND *dr1* is big
tL3_left big AND *tL3_right* big	*crn2* is zero AND *dr1* is big

Fig. 3.30. Fuzzy rules of OFFIS-04

The TPO decision module calculates the maximum TPO probability as follows:

$$wall = gr1 \cdot wll, \tag{3.38}$$

$$corner = \max((gr1 \cdot crn1), (gr2 \cdot crn2)), \tag{3.39}$$

$$door = \max((gr2 \cdot dr1), (gr3 \cdot dr2)), \tag{3.40}$$

$$corridor = gr3 \cdot crdr, \tag{3.41}$$

$$S_{\mathrm{L}} = \max([wall, corner, door, corridor]). \tag{3.42}$$

The output S_{L} of the TPO type interpretation represents a *TPO-type* as the result of indoor object recognition. This object recognition is able to be upgraded for the Pyramid approach with higher levels. In other words, if the highest layer of the Pyramid is n, substitute **tiltL2** and **tiltL3** with **tiltL(n-1)** and **tiltL(n)**. Moreover, for future works, if the number of TPO types is m>4, the number of sub-modules of the fine classifier is (m-1) for distinguishing (m-1) pairs of TPOs.

Fuzzy rules of OFFIS-05

IF	*THEN*
tL3_left v-small OR *tL3_right* v-small	*dr2* is zero AND *crdr* is zero
tL3_left small AND *tL3_right* small	*dr2* is big AND *crdr* is zero
tL3_left small AND *tL3_right* medium	*dr2* is big AND *crdr* is zero
tL3_left small AND *tL3_right* big	*dr2* is big AND *crdr* is small
tL3_left medium AND *tL3_right* small	*dr2* is big AND *crdr* is zero
tL3_left medium AND *tL3_right* medium	*dr2* is big AND *crdr* is small
tL3_left medium AND *tL3_right* big	*dr2* is big AND *crdr* is small
tL3_left big AND *tL3_right* small	*dr2* is big AND *crdr* is small
tL3_left big AND *tL3_right* medium	*dr2* is big AND *crdr* is small
tL3_left big AND *tL3_right* big	*dr2* is small AND *crdr* is big

Fig. 3.31. Fuzzy rules of OFFIS-05

3.9.2. Obstacle direction interpretation

Consider the situation shown in Fig. 3.32. Humans can easily identify an obstacle on their left hand side, despite lacking distance information. Insects, however, are able to identify the position of the obstacle based only on optical flow. How does this work?

By comparing the layer-1 quadrants of the matrix $\mathbf{\Omega}^{\mathrm{I}}$ to each other, it is obvious that, quadrant Q_3 is the closest, because this quadrant covers the left obstacle, making the mean metrics μ_{U} and μ_{V} of this quadrant bigger than those of the other layer-1 quadrants.

Zoom in to the closest quadrant and compare the layer-2 quadrants in the matrix $\mathbf{\Omega}^{\mathrm{II}}$, belonging to the closest quadrant Q_3. That the layer-2 quadrants Q_{31} and Q_{33} are closest to each other can be seen easily.

Symbolizing the closest layer-2 quadrants with 1, and the others with 0, the matrix of obstacle direction is defined

$$\mathbf{obstacleDirection} = \begin{bmatrix} 0 & 0 & 0 & 0 \\ 0 & 0 & 0 & 0 \\ 1 & 0 & 0 & 0 \\ 1 & 0 & 0 & 0 \end{bmatrix}. \tag{3.43}$$

Compared to the classic approaches of obstacle direction interpretation (Coombs and Roberts, 1993) and (Zufferey and Floreano, 2006), this approach requires *no metric information* for the qualitative obstacle direction interpretation. Moreover, this approach identifies obstacle direction in a 2D matrix, including horizontal direction, i.e. left or right, as the classical approaches and vertical direction, i.e. upper or lower. For this reason, equation (3.43) generally applies to the general collision avoidance of visual mobile robots.

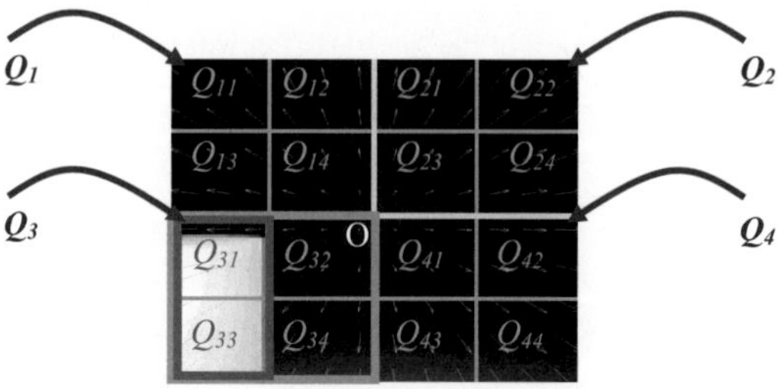

Fig. 3.32. Situation coping with an obstacle

In the scope of this work, a ground-based visual mobile robot is under consideration, so that only the horizontal direction of an obstacle is required for motion control. That means the equation (3.43) is simplified by picking up maximum values in each column. The obstacle direction is defined

$$\mathbf{dir}_{\mathrm{O}} = \begin{bmatrix} dir_{\mathrm{O}1} & dir_{\mathrm{O}2} & dir_{\mathrm{O}3} & dir_{\mathrm{O}4} \end{bmatrix}, \tag{3.44}$$

where $dir_{\mathrm{O}1} = 1$, $dir_{\mathrm{O}2} = 0$, $dir_{\mathrm{O}3} = 0$ and $dir_{\mathrm{O}4} = 0$.

3.9.3. Distance interpretation

By modifying the matrix of layer-2 quadrants $\mathbf{\Omega}^{\mathrm{II}}$ by replacing Q_{ij} with $d_{\mathrm{F}ij}$, where $d_{\mathrm{F}ij}$ is a fuzzy distance of layer-2 quadrant Q_{ij}, $d_{\mathrm{F}ij}$ is computed by the module of fuzzy distance calculation, described in section 3.7. The matrix of fuzzy distances is formed

$$\mathbf{d}_{\Omega^{\mathrm{II}}} = \begin{bmatrix} d_{\mathrm{F}11} & d_{\mathrm{F}12} & d_{\mathrm{F}21} & d_{\mathrm{F}22} \\ d_{\mathrm{F}13} & d_{\mathrm{F}14} & d_{\mathrm{F}23} & d_{\mathrm{F}24} \\ d_{\mathrm{F}31} & d_{\mathrm{F}32} & d_{\mathrm{F}41} & d_{\mathrm{F}42} \\ d_{\mathrm{F}33} & d_{\mathrm{F}34} & d_{\mathrm{F}43} & d_{\mathrm{F}44} \end{bmatrix}. \tag{3.45}$$

Using the result of obstacle direction interpretation mentioned above, the matrix of obstacle distance is defined

$$\mathbf{obstacleDistance} = \mathbf{obstacleDirection} * \mathbf{d}_{\Omega^{\mathrm{II}}}. \tag{3.46}$$

where A*B denotes the multiplication of each scalar component of A by the corresponding magnitude of the component vector of B.

For example, in the situation in Fig. 3.32, the matrix of obstacle distance is

$$\mathbf{obstacleDistance} = \mathbf{obstacleDirection} * \mathbf{d}_{\Omega^{\mathrm{II}}},$$

$$= \begin{bmatrix} 0 & 0 & 0 & 0 \\ 0 & 0 & 0 & 0 \\ 1 & 0 & 0 & 0 \\ 1 & 0 & 0 & 0 \end{bmatrix} * \begin{bmatrix} d_{\mathrm{F}11} & d_{\mathrm{F}12} & d_{\mathrm{F}21} & d_{\mathrm{F}22} \\ d_{\mathrm{F}13} & d_{\mathrm{F}14} & d_{\mathrm{F}23} & d_{\mathrm{F}24} \\ d_{\mathrm{F}31} & d_{\mathrm{F}32} & d_{\mathrm{F}41} & d_{\mathrm{F}42} \\ d_{\mathrm{F}33} & d_{\mathrm{F}34} & d_{\mathrm{F}43} & d_{\mathrm{F}44} \end{bmatrix}$$

$$= \begin{bmatrix} 0 & 0 & 0 & 0 \\ 0 & 0 & 0 & 0 \\ d_{\mathrm{F}31} & 0 & 0 & 0 \\ d_{\mathrm{F}33} & 0 & 0 & 0 \end{bmatrix}.$$

Like obstacle direction interpretation, the result of the obstacle distance interpretation generally applies to the motion control of visual mobile robots.

To particularize the obstacle distance interpretation for ground-based visual mobile robots, equation (3.46) is simplified by absorbing minimum values d_{Fij} in each column of the **obstacleDistance**. In this way, the matrix of obstacle distance becomes

$$\mathbf{d}_{O} = \begin{bmatrix} d_{O1} & d_{O2} & d_{O3} & d_{O4} \end{bmatrix}. \tag{3.47}$$

In the example under consideration, $d_{O1} = \min(d_{F31}, d_{F33})$, $d_{O2} = 0$, $d_{O3} = 0$ and $d_{O4} = 0$.

Unlike obstacle distance interpretation, the TPO distance, i.e. landmark distance, uses the matrix **roughShape** (review section 3.6.3) instead of the matrix **obstacleDirection**. For this reason, the TPO distance is defined

$$\mathbf{TPOdistance} = \mathbf{roughShape} * \mathbf{d}_{\Omega^{II}}. \tag{3.48}$$

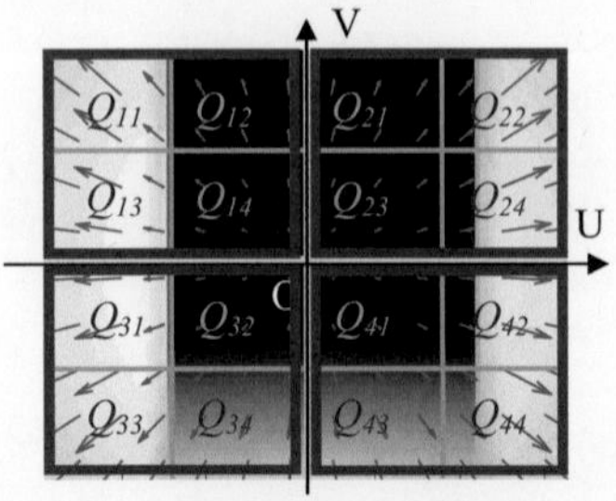

Fig. 3.33. Situation coping with a door (the same Fig 3.17)

For example, in the situation illustrated in Fig. 3.33, the same Fig. 3.17, the matrix of TPO distance is

$$\mathbf{TPOdistance} = \mathbf{roughShape} * \mathbf{d}_{\Omega^{II}},$$

$$= \begin{bmatrix} 1 & 0 & 0 & 1 \\ 1 & 0 & 0 & 1 \\ 1 & 0 & 0 & 1 \\ 1 & 0 & 0 & 1 \end{bmatrix} * \begin{bmatrix} d_{F11} & d_{F12} & d_{F21} & d_{F22} \\ d_{F13} & d_{F14} & d_{F23} & d_{F24} \\ d_{F31} & d_{F32} & d_{F41} & d_{F42} \\ d_{F33} & d_{F34} & d_{F43} & d_{F44} \end{bmatrix}$$

$$= \begin{bmatrix} d_{F11} & 0 & 0 & d_{F22} \\ d_{F13} & 0 & 0 & d_{F24} \\ d_{F31} & 0 & 0 & d_{F42} \\ d_{F33} & 0 & 0 & d_{F44} \end{bmatrix}.$$

As in the simplification of $\mathbf{d}_{Ob}$ in equation (3.47), the TPO distance is defined

$$\mathbf{d}_{L} = \begin{bmatrix} d_{L1} & d_{L2} & d_{L3} & d_{L4} \end{bmatrix}. \tag{3.49}$$

In the considered example, $d_{L1} = \min(d_{F11}, d_{F13}, d_{F31}, d_{F33})$, $d_{L2} = 0$, $d_{L3} = 0$, and $d_{L4} = \min(d_{F22}, d_{F24}, d_{F42}, d_{F44})$.

3.10. Outlier detection filter of optical flow

It is well known that the optical flow vectors contain outliers due to deficiencies in the optical flow computation, e.g. noise of the real environment. For this reason, the implementation of an outlier detection filter is necessary to improve the quality of the optical flow data before semantic interpretation of the optical flow field data. So far, noise reduction of the optical flow field in real environments is still a big challenge.

To reject optical flow outliers, (Zhang et al. 1994) proposed an approach for image processing that only exploited the epipolar geometry constraint. The central idea of this approach is in combination with feature-matching schemata to find obvious error points and epipolar geometry iteration to remove outliers. The drawbacks of this approach are, however, that a random selection scheme must be adopted and a lot of computation time for the matching of features is needed.

Alternatively, (Iu and Lin, 1999) chose a rank-based decision function using a threshold technique to categorize outliers. The disadvantages of this approach are that a constant threshold for all pixels must be used, and too much computation time is consumed for the iterations that compare predicted images with reference images.

In this thesis, a novel scheme for an optical flow outlier detection filter is presented. The key idea of the outlier detection filter is the qualitative detection of an outlier based on comparison between the optical flow vector and the average value of an optical flow pattern related to the considered vector.

Fig. 3.34. Optical flow under the ideal condition of simulation

To explain the algorithm for detecting an outlier, first consider an optical flow field under ideal conditions, e.g. the simulated image shown in Fig. 3.34. The layer-2 quadrant Q_{13} contains six optical flow vectors. The amplitudes of these vector mappings on the U-axis are given by the matrix

$$|Q_{13}|_U = \begin{bmatrix} u_{11} & u_{12} & u_{13} \\ u_{21} & u_{22} & u_{23} \end{bmatrix}, \tag{3.50}$$

By inspection the amplitudes of the optical flow vectors in each column are (under the assumption of a smooth surface) quite similar to each other, i.e. $u_{11} \approx u_{21}$, $u_{12} \approx u_{22}$ and $u_{13} \approx u_{23}$.

With respect to the three columns, the three mean values are

$$\mu_1 = \frac{(u_{11} + u_{21})}{2},\ \mu_2 = \frac{(u_{12} + u_{22})}{2} \text{ and } \mu_3 = \frac{(u_{13} + u_{23})}{2}.$$

Generally,

$$\mu_i = \frac{1}{n} \sum_{j=1}^{n} u_{ji}, \tag{3.51}$$

where

- i and j are respectively the column index and row index,
- n is the number of optical flow vectors at column i.

The ratio between the amplitudes of the maximum vector in a column and the mean value of this column is defined by

$$r_i = \frac{\max(u_{ji})}{\mu_i} \tag{3.52}$$

Generally, under ideal (non-erroneous) conditions, $r_i \approx 1$, e.g. due to $u_{13} \approx u_{23}$, $r_3 = \frac{\max(u_{13}, u_{23})}{\mu_3} \approx 1$.

Consider another optical flow image under real conditions, as illustrated in Fig. 3.35. The layer-2 quadrant Q_{13} includes 12 optical flow vectors with 4 rows and 3 columns. The amplitudes of these vector mappings on the U-axis are given by the matrix

$$|Q_{13}|_{\mathrm{U}} = \begin{bmatrix} u_{11} & u_{12} & u_{13} \\ u_{21} & u_{22} & u_{23} \\ u_{31} & u_{32} & u_{33} \\ u_{41} & u_{42} & u_{43} \end{bmatrix}. \tag{3.53}$$

By inspection it can be deduced that the element u_{43} is an optical flow outlier of this quadrant because it is too big in comparison to others in the quadrant.

Mathematically, using the equation (3.51) with $n = 4$, the three mean values of the three columns are calculated by

$$\mu_1 = \frac{\sum_{j=1}^{4} u_{j1}}{n} = \frac{(u_{11} + u_{21} + u_{31} + u_{41})}{4},$$

$$\mu_2 = \frac{\sum_{j=1}^{4} u_{j2}}{n} = \frac{(u_{12} + u_{22} + u_{32} + u_{42})}{4},$$

$$\mu_3 = \frac{\sum_{j=1}^{4} u_{j3}}{n} = \frac{(u_{13} + u_{23} + u_{33} + u_{43})}{4}.$$

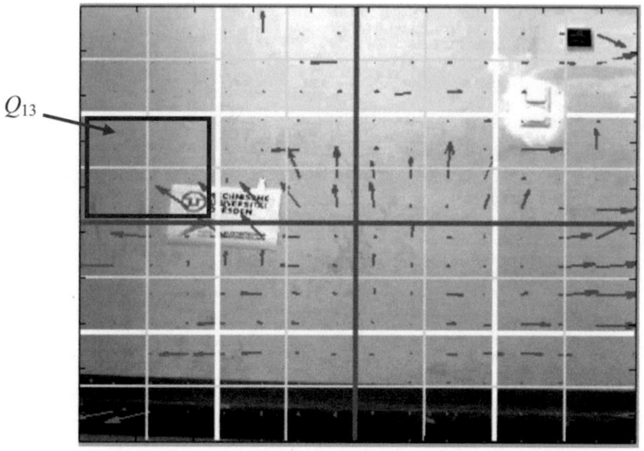

Fig. 3.35. Optical flow under the real condition of experiments

Concentrate on the mean μ_3 because this column contains the outlier u_{43}, which is also the maximum amplitude of column 3 of the layer-2 quadrant Q_{13}.

Assume that $u_{13} \approx u_{23} \approx u_{33} \approx a$, and $u_{43} = k_{\text{outlier}} \cdot a$, where a is a quantity and k_{outlier} is a multiplier coefficient, resulting in the mean value $\mu_3 = a\frac{\left(3+k_{\text{outlier}}\right)}{4}$.

Applying equation (3.52), the ratio between the maximum vector and the mean value of column 3 can be calculated as $r_3 = \frac{k_{\text{outlier}} \cdot a}{\mu_3} = 4\frac{k_{\text{outlier}}}{\left(3+k_{\text{outlier}}\right)} = n\frac{k_{\text{outlier}}}{\left((n-1)+k_{\text{outlier}}\right)}$.

It is possible to distinguish between two possible outlier situations: (i) $k_{\text{outlier}} \gg 1$ and (ii) $k_{\text{outlier}} \ll 1$. Consider the first case (i), i.e. large optical flow vector outliers.

Generally, under real conditions, if column *i* has a large outlier, then $r_i \approx n$, if $k_{\text{outlier}} \gg n$ and $n > 1$.

Extend the problem with the number of outliers to equal m, $m < n$, then the mean value of column *i* becomes $\mu_i = a\frac{\left(n-m+m \cdot k_{\text{outlier}}\right)}{n}$.

Upon applying equation (3.52) again, the ratio between the maximum vector and the mean value of column 3 can be calculated as $r_i = n\frac{k_{\text{outlier}}}{\left(n-m+m \cdot k_{\text{outlier}}\right)} = \frac{n}{m}\frac{m \cdot k_{\text{outlier}}}{\left(n-m+m \cdot k_{\text{outlier}}\right)}$.

The maximum vector is defined as an outlier if $r_i \approx \frac{n}{m}$, because $m \cdot k_{\text{outlier}} \gg (n-m)$.

In case of sufficiently few outliers, say $m < \frac{n}{2} \Rightarrow \frac{n}{m} > 2$, the value of r_i will be considerably larger than one and will mask an outlier situation.

In a critical situation with many outliers, i.e. $m > \frac{n}{2} \Rightarrow \frac{n}{m} < 2$, the r_i would be about one and will falsely indicate the situation to be a non-outlier situation, the erroneous optical flow field will not be detected as erroneous, and it will be used for further interpretation. Such a situation could only be detected if the historical data could be taken into account, which is not the case in the current implementation.

The functional principle of the outlier detection filter can be summarized as follows: if in a quadrant any optical flow vector of a column is much bigger than the average optical flow vector of the column, it will be determined as an outlier.

From the experimental analysis a sufficient optical flow quality and proper outlier detection performance is given for $\frac{n}{m} \geq 3$, i.e. the number of outliers is not bigger than one-third of number of total vectors in the considered area, because the outlier ratio $r_i \approx \frac{n}{m}$ is sufficiently separated from the non-outlier ratio $r_i \approx 1$.

A much simpler analysis can be applied for outlier case (ii) $k_{\text{outlier}} \ll 1$, i.e. small sized outliers. In such a case the outlier vector is interpreted as a non-outlier vector because the subsequent interpretations only pay attention to the averaged metrics, which means the error induced by the small outliers is tolerable.

The same procedure of outlier detection as applied to the U-axis projections of optical flow vectors can be applied to the V-axis projections by just analyzing the rows of the amplitude matrix instead of the columns.

In actuality, an optical flow quadrant may contain non-flat surface. That means the amplitudes of the optical flow vectors in each column and/or each row are not similar to each other. For this reason, to improve the quality of the outlier detection filter, it is also possible to apply the layer-3 quadrants (or layer-n quadrants, n>2) instead of the layer-2 quadrants, and compute the average value of a layer-3 quadrant instead of the average value of a column.

In conclusion, the proposed filter determines the outliers by comparing the amplitude of an optical flow vector with the averaged amplitude of an optical flow pattern related to the considered vector. Notice that although this approach also uses a threshold technique, the outlier threshold is not constant but changes with respect to a considered optical flow pattern. This makes the outlier detection more flexible and robust in different noisy areas.

Chapter 4
Fuzzy-based motion control (MOFIS)

4.1. Overview

In the recent decades, the motion control for mobile robots under changing and uncertain environments has been a challenging problem. To solve this problem, behavior-based control has recently emerged as one of the preferred solutions because it gives mobile robots the reasonable flexibility to adapt to changes and uncertainty of environments and the intelligent ability to manage multi-objective tasks.

Moreover, behavior-based control provides mobile robots with efficient solutions based on a collection of self-contained behaviors, called primitive behaviors. A primitive behavior is designed to carry out a single control policy or goal of which the priority might be changed. To operate primitive behaviors simultaneously and/or sequentially with respect to different objectives, behavior-based control needs a rational architecture to coordinate them.

The behavior-based control architecture of the MOFIS behaviors is organized as shown in Fig. 4.1 similar to the hierarchical architecture introduced by (Tunstel, 1996). In this hierarchical structure, the MOFIS behaviors are decomposed into a bottom-up hierarchy of increased behavioral complexity. A behavior at a given level is dependent on behaviors at the level below. The primitive behaviors reside at the lowest level.

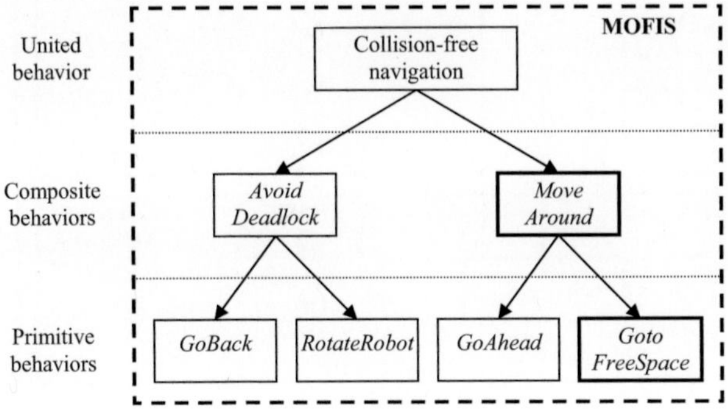

Fig. 4.1. Hierarchical architecture of the MOFIS behaviors

The MOFIS module, illustrated in Fig. 4.2, executes four main tasks: primitive behavior calculation, composite behavior calculation, behavior selection and kinematic fusion. The first three tasks are based merely on information about obstacles, including obstacle direction $\mathbf{dir}_O$ and obstacle distance $\mathbf{d}_O$, taken from the OFFIS module (see chapter 3). The last task uses not only the local obstacle information, but also a goal-oriented angle α_G and a velocity command $\mathbf{v}_{PP}$ of the PAPFIS module (see chapter 5).

The primitive behaviors of the MOFIS are *GoBack*, *GoAhead*, *RotateRobot*, and *GotoFreeSpace* with four competitive output velocities $\mathbf{v}_{GB}$, $\mathbf{v}_{GA}$, $\mathbf{v}_{RR}$, and $\mathbf{v}_{FS}$, respectively. The composite behaviors of the MOFIS are *AvoidDeadlock* and *MoveAround* with two competitive output velocities $\mathbf{v}_{AD}$, and $\mathbf{v}_{MA}$, respectively.

The MOFIS behavior selector executes a competitive coordination between the two composite behaviors. The output of this selector is the velocity command $\mathbf{v}_{MO}$ for a collision-free behavior.

The fuzzy-based kinematic fusion performs a cooperative coordination between the collision-free behavior and the goal-oriented behavior (described in chapter 5), i.e. a fusion of the two competitive velocity commands $\mathbf{v}_{MO}$ and $\mathbf{v}_{PP}$.

The output of the fusion is a *set-point*, $\mathbf{v} = [v, \omega]$, that commands the locomotion of the autonomous mobile robot (AMR).

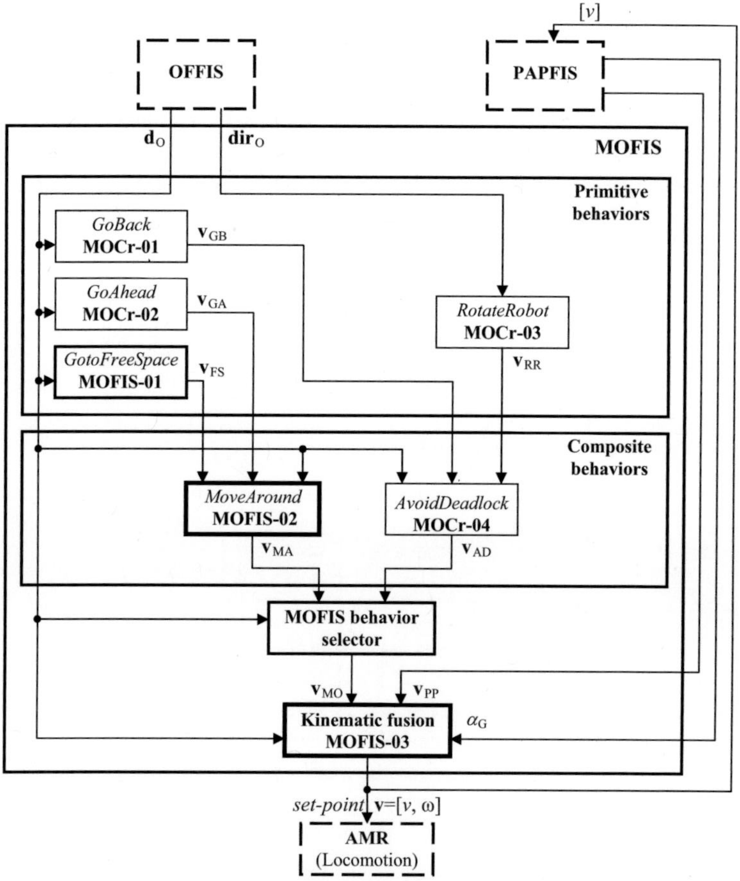

Fig. 4.2. Organization of the MOFIS module

In the following sections of this chapter, the description of mobile robot kinematics is introduced. Then, the basics of behavior-based robotics, including behavior features and behavioral coordination, are briefly described. After that, the primitive and composite behaviors of MOFIS are specifically discussed. Some of these behaviors use fuzzy logic concepts, which make robotic motions as flexible and smooth as biological systems. Finally, the MOFIS behavior selector and fuzzy-based kinematic fusion will be presented.

4.2. Description of differential drive mobile robot kinematics

4.2.1. Kinematics description in the inertial frame

When analyzing the kinematic properties, known as the differential drive, assume a mobile robot is a rigid body on two drive wheels mounted on a common axis, and its wheels can independently be driven either forward or backward on flat surfaces. Under this assumption, the pose of the robot chassis on the flat surface can be represented by three coordinates, two of which (x, y) represent the positions of the robot on the plane ∠(XOY), and the last of which θ represents the orientation of the robot with respect to the vertical Z axis orthogonal to the plane ∠(XOY), illustrated in Fig. 4.3.

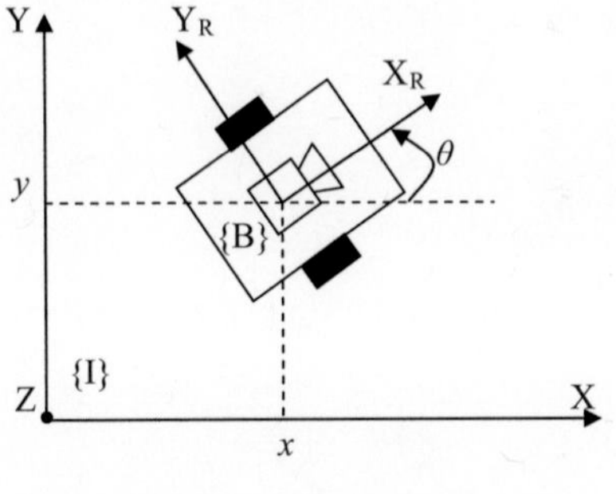

Fig. 4.3. Kinematic description in the inertial frame

Specify two reference frames, including the inertial reference frame {I} and body frame {B}, in order to represent the coordinates (x, y, θ).

The inertial reference frame {I} is a global reference frame to specify the position of the robot chassis through the coordinate system (XYZ). This reference frame is assumed to be stationary. The position (x, y) of the robot is represented in this reference frame.

The body reference frame {B} is a local reference frame, represented by the coordinate system ($X_R Y_R Z_R$), which is fixed to the robot. This frame defines any difference in orientation between the two frames. The orientation θ of the robot is represented by a rotational angle from the X-axis to the X_R-axis around the Z-axis.

The kinematic description of the robot in the inertial frame can be defined by

$$\begin{bmatrix}\dot{x}\\ \dot{y}\\ \dot{\theta}\end{bmatrix} = \mathbf{R}_z(\theta)\cdot\begin{bmatrix}v\\ \omega\end{bmatrix} = \begin{bmatrix}\cos\theta & -\sin\theta & 0\\ \sin\theta & \cos\theta & 0\\ 0 & 0 & 1\end{bmatrix}\cdot\begin{bmatrix}v\\ \omega\end{bmatrix}, \tag{4.1}$$

where

- v and ω are the linear and angular velocities in the frame {B},
- $\dot{x}$ and $\dot{y}$ are the linear velocities on the X-axis and Y-axis,
- θ is the rotational angle of the frame {B} related to the frame {I},
- $\dot{\theta}$ is the angular velocity on the Z-axis of the frame {I},
- $\mathbf{R}_z(\theta) = \begin{bmatrix}\cos\theta & -\sin\theta & 0\\ \sin\theta & \cos\theta & 0\\ 0 & 0 & 1\end{bmatrix}$ is the rotation matrix on the Z-axis.

4.2.2. Kinematics description in the polar coordinates

With the assumption that a robot moves from a position $[x, y, \theta]^{\mathrm{T}}$ to a goal $\mathbf{P}_{\mathrm{G}} = [x_{\mathrm{G}}, y_{\mathrm{G}}, \theta_{\mathrm{G}}]^{\mathrm{T}}$. The difference distance between the two positions is Δd with respect to the two differences Δx and Δy on axes X and Y, respectively. The goal-oriented angle α_{G}, and the goal direction θ_{G} are illustrated in Fig. 4.4.

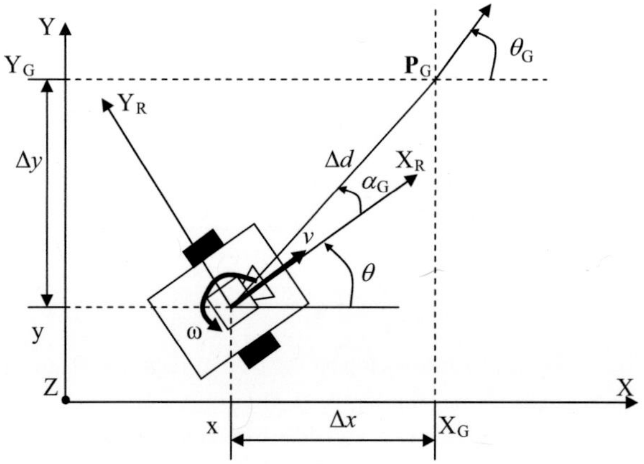

Fig. 4.4. Kinematic description in polar coordinates

The coordinate transformation into the polar coordinates with its origin at the goal position can be performed by

$$\Delta d = \sqrt{\left(\Delta x^2 + \Delta y^2\right)}, \tag{4.2}$$

$$\alpha_{\mathrm{G}} = \operatorname{atan2}\left(\Delta y, \Delta x\right) - \theta, \tag{4.3}$$

$$\theta_{\mathrm{G}} = \theta + \alpha_{\mathrm{G}}. \tag{4.4}$$

The kinematics description in polar coordinates can be defined by

$$\begin{bmatrix} \Delta \dot{d} \\ \dot{\alpha}_{\mathrm{G}} \\ \dot{\theta}_{\mathrm{G}} \end{bmatrix} = \begin{bmatrix} \cos \alpha_{\mathrm{G}} & 0 \\ \dfrac{\sin \alpha_{\mathrm{G}}}{\delta} & -1 \\ \dfrac{\sin \alpha_{\mathrm{G}}}{\delta} & 0 \end{bmatrix} \cdot \begin{bmatrix} v \\ \omega \end{bmatrix} \text{ for } \alpha_{\mathrm{G}} \in \left[-\frac{\pi}{2}, \frac{\pi}{2}\right], \tag{4.5}$$

and

$$\begin{bmatrix} \Delta \dot{d} \\ \dot{\alpha}_{\mathrm{G}} \\ \dot{\theta}_{\mathrm{G}} \end{bmatrix} = \begin{bmatrix} \cos \alpha_{\mathrm{G}} & 0 \\ -\dfrac{\sin \alpha_{\mathrm{G}}}{\delta} & -1 \\ -\dfrac{\sin \alpha_{\mathrm{G}}}{\delta} & 0 \end{bmatrix} \cdot \begin{bmatrix} v \\ \omega \end{bmatrix} \text{ for } \alpha_{\mathrm{G}} \in \left[-\pi, -\frac{\pi}{2}\right) \cup \left(\frac{\pi}{2}, \pi\right]. \tag{4.6}$$

Further details concerning the basics of differential drive mobile robot kinematics are described in (Berns and Puttkamer, 2009).

4.3. Basics of behavior-based robotics

4.3.1. Definition of notation

Consider an example with the assumption that a robot can distinguish three physical forms of obstacles, doors, and paths. Define notations as the following:

- $\mathbf{S}$ denotes a vector of all stimuli $\mathbf{s}_i$, $i = 1 \ldots 3$ with respect to three predefined stimuli,

$$\mathbf{S} = \begin{bmatrix} \mathbf{s}_1 \\ \mathbf{s}_2 \\ \mathbf{s}_3 \end{bmatrix} = \begin{bmatrix} (p_1, \lambda_1) \\ (p_2, \lambda_2) \\ (p_3, \lambda_3) \end{bmatrix} = \begin{bmatrix} (\text{detected-obstacle}, 0.4) \\ (\text{detected-door}, 0.9) \\ (\text{detected-path}, 0.1) \end{bmatrix}. \tag{4.7}$$

In this example, the robot detected an obstacle in a medium distance with the stimulus strength 0.4, a door in a near distance with the stimulus strength 0.9, and a path in a very far distance with the stimulus strength 0.1.

- $\mathbf{B}$ denotes a vector of all active behaviors β_i at a given time t,

$$\mathbf{B}(\mathbf{S}) = \begin{bmatrix} \beta_1(\mathbf{s}_1) \\ \beta_2(\mathbf{s}_2) \\ \beta_3(\mathbf{s}_3) \end{bmatrix} = \begin{bmatrix} \beta_{\text{avoid-obstacle}}(\mathbf{s}_1) \\ \beta_{\text{go-to-door}}(\mathbf{s}_2) \\ \beta_{\text{follow-path}}(\mathbf{s}_3) \end{bmatrix}. \tag{4.8}$$

- $\mathbf{R}$ denotes a vector of all responses $\mathbf{r}_i$, $\mathbf{r}_i = [x_i, y_i, \theta_i]$ for flat ground-based mobile robots. $\mathbf{R}$ is computed by the set of active behaviors $\beta : S \to R$,

$$\mathbf{R} = \begin{bmatrix} \mathbf{r}_1 \\ \mathbf{r}_2 \\ \mathbf{r}_3 \end{bmatrix} = \begin{bmatrix} \mathbf{r}_{\text{avoid-obstacle}} \\ \mathbf{r}_{\text{go-to-door}} \\ \mathbf{r}_{\text{follow-path}} \end{bmatrix} \tag{4.9}$$

with the component vector magnitudes equal to (*arbitrarily* for this case)

$$\mathbf{R}_m = \begin{bmatrix} 0.6 \\ 1.0 \\ 0.0 \end{bmatrix}.$$

- $\mathbf{G}$ denotes a vector encoding the relative strength or gain g_i of each active behavior β_i,

$$\mathbf{G} = \begin{bmatrix} g_1 \\ g_2 \\ g_3 \end{bmatrix} = \begin{bmatrix} g_{\text{avoid-obstacle}} \\ g_{\text{go-to-door}} \\ g_{\text{follow-path}} \end{bmatrix} = \begin{bmatrix} 1.5 \\ 1.2 \\ 0.8 \end{bmatrix}. \tag{4.10}$$

- C denotes a behavioral coordination function. The function is now defined by

$$\zeta = \mathrm{C}(\mathbf{R}') \tag{4.11}$$

where $\mathbf{R}' = \mathbf{G} * \mathbf{R}$, and * denotes the multiplication of each scalar component (g_i) by the corresponding magnitude of the component vector ($\mathbf{r}_i$). The function is rewritten by

$$\zeta = \mathrm{C}\left(\begin{bmatrix} g_1 * \mathbf{r}_1 \\ g_2 * \mathbf{r}_2 \\ g_3 * \mathbf{r}_3 \end{bmatrix}\right) \tag{4.12}$$

with scaled component vector magnitudes

$$\mathbf{R}'_m = \mathbf{G} * \mathbf{R}_\mathrm{m} = \begin{bmatrix} 1.5 * 0.6 \\ 1.2 * 1.0 \\ 0.8 * 0.0 \end{bmatrix} = \begin{bmatrix} 0.9 \\ 1.2 \\ 0.0 \end{bmatrix}.$$

$\mathbf{G}$ is only valuable when multiple behaviors are coordinated, enabling the robot to set priorities through establishing the relative importance of each of its constituent behaviors.

By using a winner-take-all coordination strategy, a single component vector $g_i * \mathbf{r}_i$ with the greatest magnitude will be chosen in $\mathbf{R}'_m$. In the example, the *go-to-door* behavior, i.e. the second element of $\mathbf{R}'_m$, with the component value of 1.2 is selected.

By using a high-priority-take-all coordination strategy, the highest-ranked component behavior encoded in $\mathbf{G}$ would be chosen and activated. This choice is independent of the component vector magnitudes in $\mathbf{R}'_m$. In the example, the *avoid-obstacle* behavior, i.e. the first element of $\mathbf{G}$, with the highest priority $g_1 = 1.5$ is activated.

Generally, with *n* stimuli we obtain

$$\mathbf{S} = \begin{bmatrix} \mathbf{s}_1 \\ \vdots \\ \mathbf{s}_n \end{bmatrix}, \ \mathbf{B}(\mathbf{S}) = \begin{bmatrix} \beta_1(\mathbf{s}_1) \\ \vdots \\ \beta_n(\mathbf{s}_n) \end{bmatrix}, \ \mathbf{R} = \begin{bmatrix} \mathbf{r}_1 \\ \vdots \\ \mathbf{r}_n \end{bmatrix}, \ \mathbf{G} = \begin{bmatrix} g_1 \\ \vdots \\ g_n \end{bmatrix}, \text{ and } \zeta = \mathrm{C}\left(\begin{bmatrix} g_1 * \mathbf{r}_1 \\ \vdots \\ g_n * \mathbf{r}_n \end{bmatrix}\right).$$

Furthermore, a behavioral coordination function can recursively be referred to as an input of another coordination function. For this reason, the coordination function can be defined as

$$\zeta' = \mathrm{C}'\left(\begin{bmatrix} \zeta_1 \\ \vdots \\ \zeta_n \end{bmatrix}\right), \tag{4.13}$$

where C' is a behavioral coordination function.

In the way of behavioral coordination, a robotic system can deal with multiple behaviors that are concurrently activated. That means the robotic system implements the behavioral coordination function C to solve the conflict between the behaviors.

The behavioral coordination functions are classified into two coordination methods: competitive coordination and cooperative coordination.

4.3.2. Behavioral coordination methods

a) Competitive methods

Competitive methods supply the means for coordinating behavioral responses to solve the conflict that stems from activating multiple behaviors concurrently.

The coordinator is usually considered to be a winner-take-all network where the single response for the winning behavior is chosen and executed, see Fig. 4.5. Based on various criteria, the competitive coordinator decides which behavior is best for the robot control.

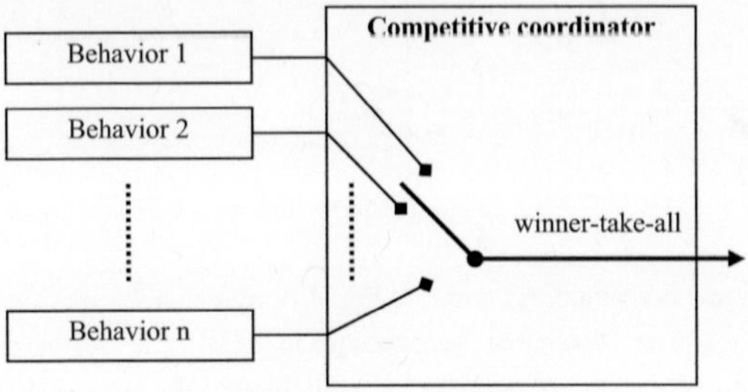

Fig. 4.5. Competitive coordinator

Basically, two main methods for realizing competitive methods are subsumption-based methods (Brooks, 1986) and action-selection methods (Mayes, 1990).

"Subsumption-based methods use an arbitrative coordination function that has taken the form of a fixed prioritization network in which a strict behavioral dominance hierarchy exists, typically through the use of suppression and inhibition in a descending manner" (Arkin, 1999).

Action-based methods have no fixed hierarchy. They arbitrarily select the output of a single behavior based on the competition of all active behaviors. This coordination is done "in a less autocratic manner" (Arkin, 1999) in comparison with subsumption-based methods. The competition of active behaviors is done based on the activation levels guided by the agent's goals and the perception information.

Generally, a competitive coordinator only allows one behavior to be activated at any given moment. Therefore, the important property of the competitive coordination is *robustness*. For example, in coping with a danger only a safety behavior is selected to avoid the danger. Additionally, the competitive coordination allows for the *easy adaption* because the final output only depends on the particular behavior at any given moment.

b) Cooperative methods

Cooperative methods support the coordination functions with behavioral fusion to activate *more than one* behavior at a time. They can be enacted in a variety of ways, such as, potential-fields approaches (Khatib, 1985) and behavioral blending approaches (Saffiotti et al. 1995).

In the potential-fields approach, each behavior's reactive strength or gain is a multiplier of the vectors before addition. In another way, in the behavioral blending approach "each behavior is assigned a desirability function that can be combined using specialized logical operators called t-norms, a form of multi-valued logic" (Arkin, 1999).

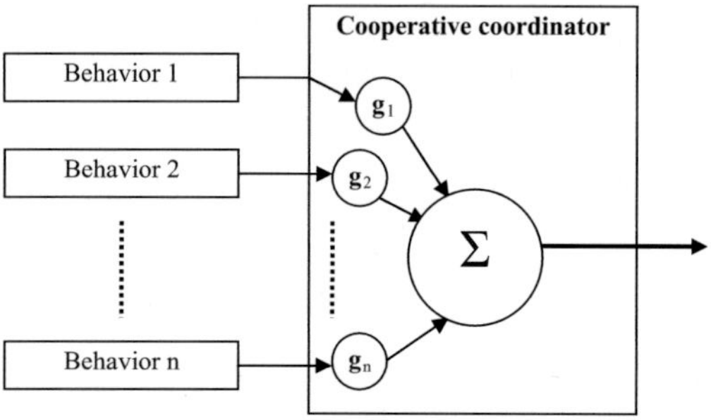

Fig. 4.6. Cooperative coordinator

The cooperative coordinator, shown in Fig. 4.6, makes a fusion of the input behaviors by a summation to find an amenable representation. The key property of this cooperative coordinator is enabling *smoothness* because all behaviors are activated at a moment. This means the output response will always be that all behaviors will merge. Additionally, the smoothness of the output response is easy to be tuned by changing the strengths of the input behaviors, i.e. adjusting gain factors $\mathbf{g}_i$ of the summation.

4.4. MOFIS primitive behaviors

According to (Arkin, 1999), reactive control is a technique for tightly coupling perception and action, typically in the context of motor behaviors, to produce timely robotic responses in dynamic and unstructured worlds. That means the advantage of reactive behaviors is the speed of the response between robots and their environment.

The primitive behaviors of the MOFIS are reactive behaviors performing an individual duty by operating in a reactive paradigm. The primitive behaviors of the MOFIS are *GoBack*, *GoAhead*, *RotateRobot* and *GotoFreeSpace*.

The three first primitive behaviors are monotonic because their motion direction is unchanged and independent of the stimulus strength. These behaviors use crisp data, so that they are symbolized with MOCr-01, 02 and 03, see Fig. 4.2.

The primitive behavior *GoBack* is responsible for controlling a mobile robot going backward. The velocity command of this behavior has the form

$$\mathbf{v}_{GB} = \left[v_{GB}, \omega_{GB} \right], \tag{4.14}$$

where

- v_{GB} is the linear velocity, $v_{GB} \in \left[-1 \quad 0 \right]$, $v_{GB} = -1 + d_{Os}$, $d_{Os}=\min(\mathbf{d}_O)$,
- ω_{GB} is the angular velocity, $\omega_{GB} = 0$.
- $\mathbf{d}_O = \left[d_{O1} \quad d_{O2} \quad d_{O3} \quad d_{O4} \right]$ is the obstacle distance defined in chapter 3.

In contrast with the behavior *GoBack*, the behavior *GoAhead* drives a mobile robot forward. The velocity command of the behavior *GoAhead* has the form

$$\mathbf{v}_{GA} = \left[v_{GA}, \omega_{GA} \right], \tag{4.15}$$

where $\omega_{GA} = 0$ and $v_{GA} = d_{Os}=\min(\mathbf{d}_O)$, $v_{GA} \in \left[0 \quad 1 \right]$.

In another way, the primitive behavior *RotateRobot* controls the rotation of a mobile robot following a direction opposite to the direction $\mathbf{dir}_O$. For this reason, the velocity command of this behavior has the form

$$\mathbf{v}_{RR} = \left[v_{RR}, \omega_{RR} \right], \tag{4.16}$$

where

- $v_{RR} = 0$,
- $\omega_{RR} = \begin{cases} 1 & \Leftrightarrow \textit{rotateRight} \text{ if } \max(\mathbf{dir}_O) = dir_{O1} \text{ OR } dir_{O2} \\ \text{-}1 & \Leftrightarrow \textit{rotateLeft} \text{ if } \max(\mathbf{dir}_O) = dir_{O3} \text{ OR } dir_{O4} \end{cases}$,
- $\mathbf{dir}_O = \left[dir_{O1} \quad dir_{O2} \quad dir_{O3} \quad dir_{O4} \right]$ is obstacle direction defined in chapter 3.

Notice that the rotation direction of the primitive behavior *RotateRobot* is kept until the deadlock avoidance is finished.

Also using obstacle distance $\mathbf{d}_O$ and obstacle direction $\mathbf{dir}_O$ as the inputs, the primitive behavior *GotoFreeSpace* applies a fuzzy inference system to compute its velocity command. The module MOFIS-01, illustrated in Fig. 4.7, is designed for the behavior *GotoFreeSpace*.

The sub-module for parameter calculation calculates the three inputs of the fuzzy inference. These parameters are deduced from obstacle distance $\mathbf{d}_{Ob}$, including

- $dr_{\text{Ob}} = \begin{cases} left & \text{if } \min(\mathbf{d}_{\text{O}}) = d_{\text{O1}} \text{ OR } d_{\text{O2}} \\ right & \text{if } \min(\mathbf{d}_{\text{O}}) = d_{\text{O3}} \text{ OR } d_{\text{O4}} \end{cases}$,
- $d_{\text{Ob12}} = \min(d_{\text{O1}}, d_{\text{O2}})$,
- $d_{\text{Ob34}} = \min(d_{\text{O3}}, d_{\text{O4}})$.

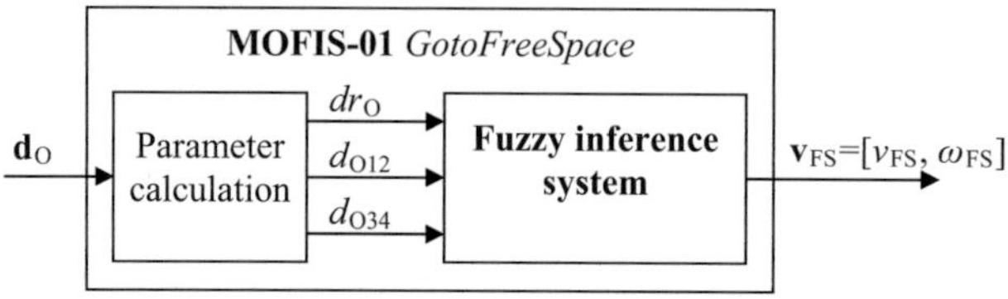

Fig. 4.7. Fuzzy-based module of the primitive behavior *GotoFreeSpace*

Fuzzy rules of MOFIS-01

IF	*THEN*
dr_{Ob}= *left* AND d_{Ob12}= *v-near* AND d_{Ob34}= *v-near*	v_{FS} = *zero* AND ω_{FS} = *zero*
dr_{Ob}= *left* AND d_{Ob12}= *v-near* AND d_{Ob34}= *near*	v_{FS} = *zero* AND ω_{FS} = *zero*
dr_{Ob}= *left* AND d_{Ob12}= *v-near* AND d_{Ob34}= *mid*	v_{FS} = *zero* AND ω_{FS} = *smallPos*
dr_{Ob}= *left* AND d_{Ob12}= *v-near* AND d_{Ob34}= *far*	v_{FS} = *zero* AND ω_{FS} = *midPos*
dr_{Ob}= *left* AND d_{Ob12}= *v-near* AND d_{Ob34}= *v-far*	v_{FS} = *zero* AND ω_{FS} = *bigPos*
dr_{Ob}= *left* AND d_{Ob12}= *near* AND d_{Ob34}= *near*	v_{FS} = *slow* AND ω_{FS} = *zero*
dr_{Ob}= *left* AND d_{Ob12}= *near* AND d_{Ob34}= *mid*	v_{FS} = *slow* AND ω_{FS} = *smallPos*
dr_{Ob}= *left* AND d_{Ob12}= *near* AND d_{Ob34}= *far*	v_{FS} = *slow* AND ω_{FS} = *midPos*
dr_{Ob}= *left* AND d_{Ob12}= *near* AND d_{Ob34}= *v-far*	v_{FS} = *slow* AND ω_{FS} = bigPos
dr_{Ob}= *left* AND d_{Ob12}= *mid* AND d_{Ob34}= *mid*	v_{FS} = *med* AND ω_{FS} = *smallPos*
dr_{Ob}= *left* AND d_{Ob12}= *mid* AND d_{Ob34}= *far*	v_{FS} = *med* AND ω_{FS} = *midPos*
dr_{Ob}= *left* AND d_{Ob12}= *mid* AND d_{Ob34}= *v-far*	v_{FS} = *med* AND ω_{FS} = *bigPos*
dr_{Ob}= *left* AND d_{Ob12}= *far* OR d_{Ob12}= *v-far*	v_{FS} = *fast* AND ω_{FS} = *zero*
dr_{Ob}= *right* AND d_{Ob12}= *v-near* AND d_{Ob34}=*v-near*	v_{FS} = *zero* AND ω_{FS} = *zero*
dr_{Ob}= *right* AND d_{Ob12}= *near* AND d_{Ob34}=*v-near*	v_{FS} = *zero* AND ω_{FS} = *zero*
dr_{Ob}= *right* AND d_{Ob12}= *mid* AND d_{Ob34}=*v-near*	v_{FS} = *zero* AND ω_{FS} = *smallNeg*
dr_{Ob}= *right* AND d_{Ob12}= *far* AND d_{Ob34}=*v-near*	v_{FS} = *zero* AND ω_{FS} = *midNeg*
dr_{Ob}= *right* AND d_{Ob12}= *v-far* AND d_{Ob34}=*v-near*	v_{FS} = *zero* AND ω_{FS} = *bigNeg*
dr_{Ob}= *right* AND d_{Ob12}= *near* AND d_{Ob34}= *near*	v_{FS} = *slow* AND ω_{FS} = *zero*
dr_{Ob}= *right* AND d_{Ob12}= *mid* AND d_{Ob34}= *near*	v_{FS} = *slow* AND ω_{FS} = *smallNeg*
dr_{Ob}= *right* AND d_{Ob12}= *far* AND d_{Ob34}= *near*	v_{FS} = *slow* AND ω_{FS} = *midNeg*
dr_{Ob}= *right* AND d_{Ob12}= *v-far* AND d_{Ob34}= *near*	v_{FS} = *slow* AND ω_{FS} = *bigNeg*
dr_{Ob}= *right* AND d_{Ob12}= *mid* AND d_{Ob34}= *mid*	v_{FS} = *med* AND ω_{FS} = *smallNeg*
dr_{Ob}= *right* AND d_{Ob12}= *far* AND d_{Ob34}= *mid*	v_{FS} = *med* AND ω_{FS} = *midNeg*
dr_{Ob}= *right* AND d_{Ob12}= *v-far* AND d_{Ob34}= *mid*	v_{FS} = *med* AND ω_{FS} = *bigNeg*
dr_{Ob}= *right* AND d_{Ob34}= *far* OR d_{Ob34}= *v-far*	v_{FS} = *fast* AND ω_{FS} = *zero*

Fig. 4.8. Fuzzy rules of MOFIS-01

The velocity command of the behavior *GotoFreeSpace* has the form

$$\mathbf{v}_{FS} = \left[v_{FS}, \omega_{FS} \right]. \quad (4.17)$$

where v_{FS} and ω_{FS} are defined by the fuzzy rules of the MOFIS-01, illustrated in Fig. 4.8.

The membership functions of dr_{Ob}, d_{Ob12} and d_{Ob34} are respectively shown in Fig. 4.9, Fig. 4.10 and Fig. 4.11.

In the linguistic expression, d_{Ob12} and d_{Ob34} belong to the linguistic set {*v-near*, *near*, *mid*, *far*, *v-far*} with respect to very near, near, middle, far and very far distances, respectively.

In a similar way of linguistic expression, v_{FS} belongs to the linguistic set {*slow*, *med*, *fast*} representing slow, medium, and fast velocities, respectively; and ω_{FS} belongs to the linguistic set {*bigNeg*, *midNeg*, *smallNeg*, *zero*, *smallPos*, *midPos*, *bigPos*} representing big negative, middle negative, small negative, zero, small positive, middle positive, and big positive angular velocities, respectively.

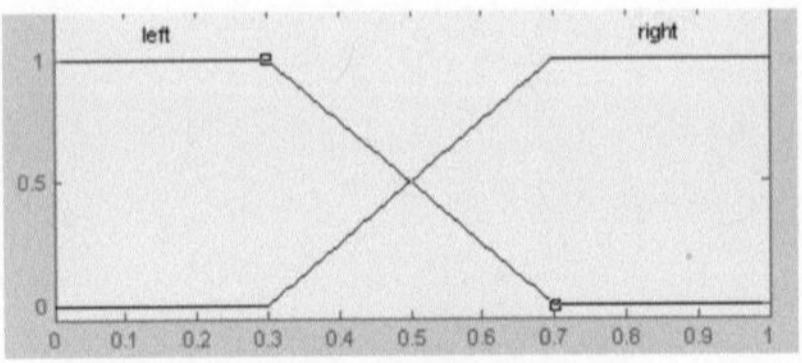

Fig. 4.9. Membership function of dr_{Ob}

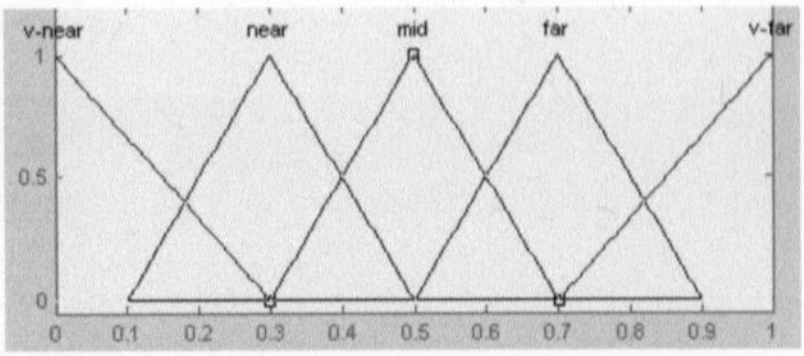

Fig. 4.10. Membership function of d_{Ob12} and d_{Ob34}

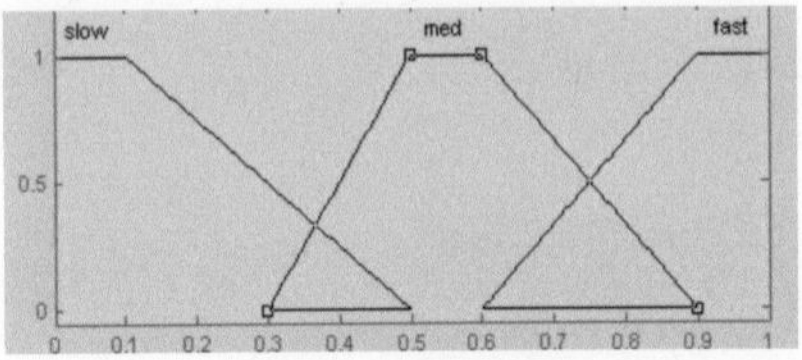

Fig. 4.11. Membership function of v_{FS}

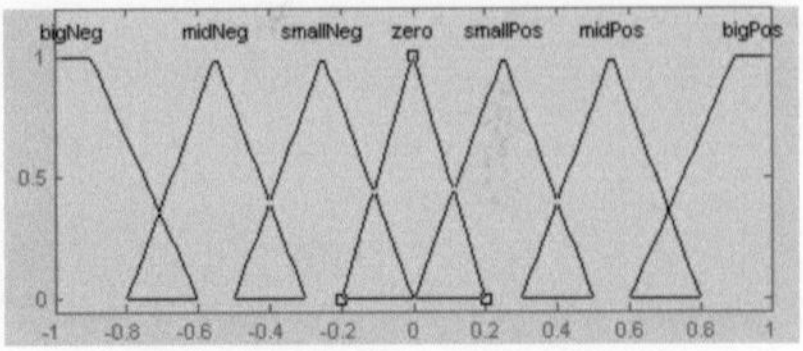

Fig. 4.12. Membership function of ω_{FS}

4.5. MOFIS composite behaviors

Not every primitive behavior is satisfactory for complex navigation tasks. However, the primitive behaviors contribute to composite behaviors to create more intelligent behaviors.

Composite behavior is a high-level behavior that performs synergic coordination between the behaviors at the lower level. Compared to the primitive behaviors, composite behaviors have better predictive capabilities to deal with changing and uncertain environments. However, the response speed of the composite behaviors is less than that of primitive behaviors (like the outer loop in conventional cascaded control).

The composite behaviors of MOFIS include *AvoidDeadlock* and *MoveAround*, displayed in Fig. 4.13.

The composite behavior *AvoidDeadlock* is responsible for driving a mobile robot avoiding a deadlock situation where a collision will certainly occur if the robot goes any further forward. It performs a deadlock avoidance strategy by arbitrating the two primitive behaviors *GoBack* and *RotateRobot* based on fuzzy obstacle distance. The velocity command of the behavior *AvoidDeadlock* has the form

$$\mathbf{v}_{\mathrm{AD}} = \left[v_{\mathrm{AD}}, \omega_{\mathrm{AD}} \right], \tag{4.18}$$

where

- $v_{\mathrm{AD}} = \left(v_{\mathrm{RR}} \cdot d_{\min} + v_{\mathrm{GB}} \cdot (1 - d_{\min}) \right)$, (4.19)
- $\omega_{\mathrm{AD}} = \left(\omega_{\mathrm{RR}} \cdot d_{\min} + \omega_{\mathrm{GB}} \cdot (1 - d_{\min}) \right)$, (4.20)
- $d_{\min} = \min(\mathbf{d}_{\mathrm{O}})$, is a minimum obstacle distance.

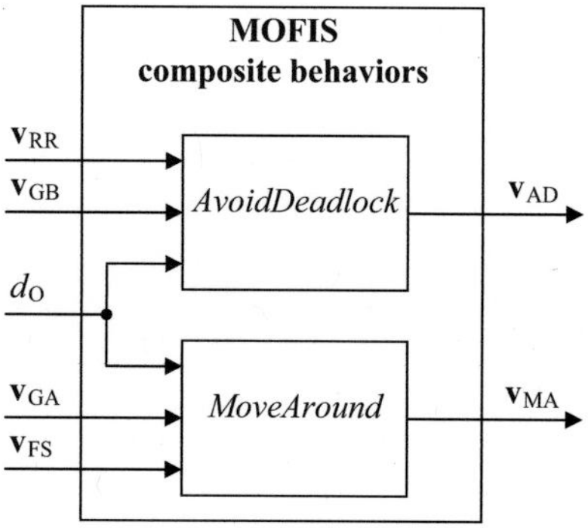

Fig. 4.13. MOFIS composite behaviors

Fuzzy rules of MOFIS-02

IF	*THEN*
$d_{\min}$ = *v-near*	k_{MA} is zero
$d_{\min}$ = *near*	k_{MA} is small
$d_{\min}$ = *mid*	k_{MA} is small
$d_{\min}$ = *far*	k_{MA} is big
$d_{\min}$ = *v-far*	k_{MA} is big

Fig. 4.14. Fuzzy rules of MOFIS-02

Unlike the behavior *AvoidDeadlock*, the composite behavior *MoveAround* has the responsibility to drive a mobile robot going forward by using fuzzy logic to collaborate two primitive behaviors *GoAhead* and *GotoFreeSpace*.

The velocity command of the behavior *MoveAround* has the form

$$\mathbf{v}_{MA} = [v_{MA}, \omega_{MA}]. \quad (4.21)$$

where

- $v_{MA} = k_{MA} \cdot v_{GA} + (1 - k_{MA}) \cdot v_{FS}$, (4.22)

- $\omega_{MA} = k_{MA} \cdot \omega_{GA} + (1 - k_{MA}) \cdot \omega_{FS}$, (4.23)

- k_{MA} is a multiplier factor resulted from d_{min} by the fuzzy rules of the module MOFIS-02, illustrated in Fig. 4.14. The membership functions of d_{min} and k_{MA} are respectively shown in Fig. 4.15 and 4.16.

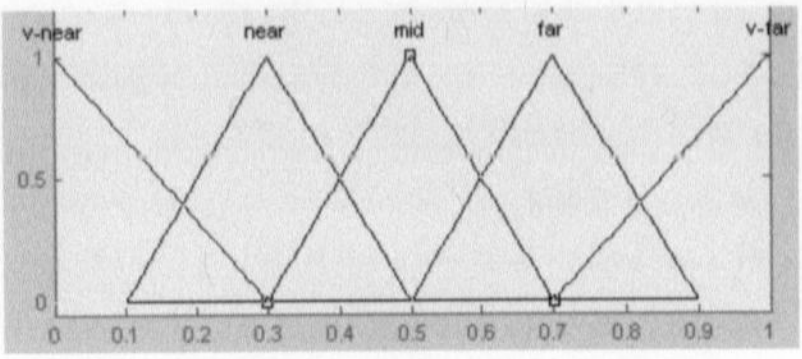

Fig. 4.16. Membership function of d_{min}

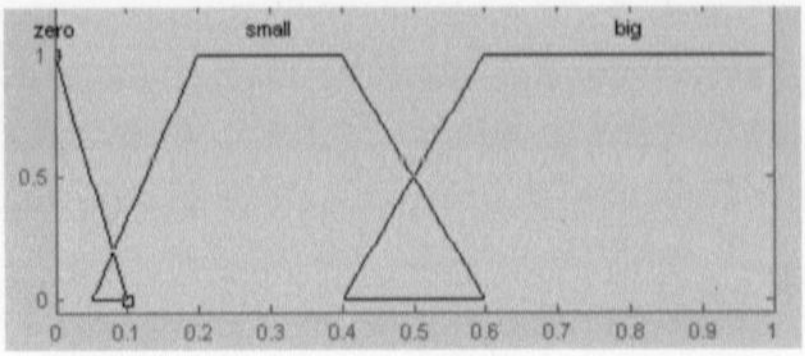

Fig. 4.17. Membership function of k_{MA}

4.6. MOFIS behavior selector

It is easy to recognize that the objectives of behaviors *AvoidDeadlock* and *MoveAround* are in conflict with each other due to the opposite motion direction. For this reason, only one of the two composite behaviors is invoked at any particular moment.

To solve this conflict, a behavioral selector is designed to activate a suitable behavior for the situation at hand. The MOFIS behavior selector, illustrated in Fig. 4.18, has three inputs: velocities $\mathbf{v}_{AD}$ and $\mathbf{v}_{MA}$ as the data signals, and obstacle distance $\mathbf{d}_O$ as the decision signal.

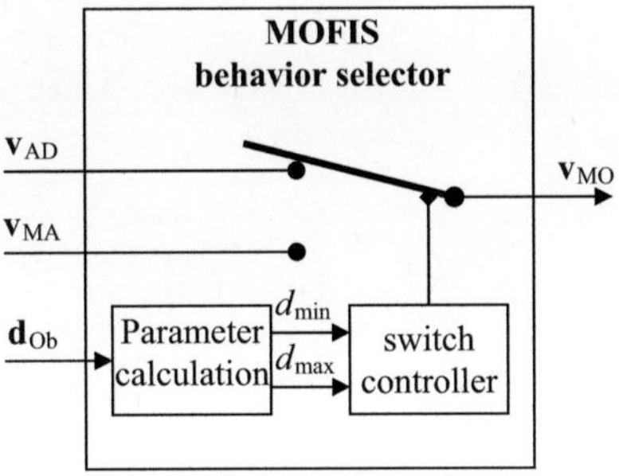

Fig. 4.18. MOFIS behavior selector

The velocity output of the MOFIS behavior selector has the form

$$\mathbf{v}_{MO} = [v_{MO}, \omega_{MO}]. \quad (4.24)$$

where:

- $v_{MO} = \begin{cases} v_{AD} & \text{if } (d_{min} \le \delta_{DL}) \vee (d_{max} \le 2 \cdot \delta_{DL}) \\ v_{MA} & \text{otherwise} \end{cases}$, (4.25)

- $\omega_{MO} = \begin{cases} \omega_{AD} & \text{if } (d_{min} \le \delta_{DL}) \vee (d_{max} \le 2 \cdot \delta_{DL}) \\ \omega_{MA} & \text{otherwise} \end{cases}$. (4.26)

- $d_{min} = \min(\mathbf{d}_O)$,
- $d_{max} = \max(\mathbf{d}_O)$,
- δ_{DL} is a dangerous radius around a deadlock defined by the user.

4.7. Fuzzy-based kinematic fusion

The objectives of the collision-free behavior and the goal-oriented behavior are not contrary to one another because they support the goal-approaching behavior at the highest level of the hierarchical structure. In other words, these behaviors are able to work together at any time. More details of the goal-oriented behavior will be described later in chapter 5.

To make a fusion of several behaviors, a large number of methods of cooperative coordination have been proposed, e.g. voting-based methods (Kittler and Alkoot, 2003) and sum-based methods (Lei, 2010). The key idea of the voting-based methods is a winner-take-all strategy enabling only one behavior to be activated at a given moment. For this reason, the voting-based methods work as a selector of component behaviors. Differently, the central idea of the sum-based methods is a summation of multiplications of active behaviors through their weights, i.e. fusion-type cooperation. These methods enable more than one behavior to be activated at a given moment. Therefore, the advantage of these state of the art methods is *the valuable ability of parallel behaviors*. However, their disadvantage is *the lack of flexibility*.

To improve the flexibility of the sum-based cooperative coordination and solve more rationally different objectives among parallel behaviors, a module of fuzzy-based kinematic fusion is proposed as illustrated in Fig. 4.19. The key idea of this kinematic fusion is the use of fuzzy logic to compute the weighting factors of the element behaviors before performing the cooperative coordination. That means a fuzzy-based cooperative coordination is performed between the MOFIS behavior and the PAPFIS behavior which is described in chapter 5. Once performed, it makes the goal-approaching behavior of mobile robots smoother.

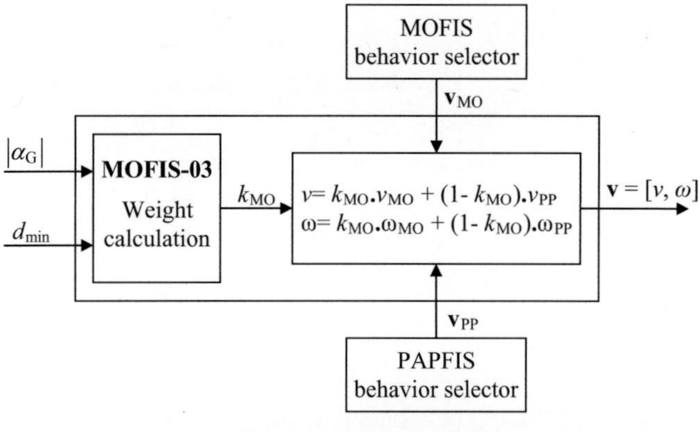

Fig. 4.19. Fuzzy-based kinematic fusion module

Fuzzy rules of MOFIS-03

IF	*THEN*
$d_{\min} = near$	k_{MO} is big
$d_{\min} \neq near$ AND $\lvert\alpha_{\mathrm{G}}\rvert = big$	k_{MO} is zero
$d_{\min} = mid$ AND $\lvert\alpha_{\mathrm{G}}\rvert = small$	k_{MO} is middle
$d_{\min} = far$ AND $\lvert\alpha_{\mathrm{G}}\rvert = small$	k_{MO} is small

Fig. 4.20. Fuzzy rules of MOFIS-03

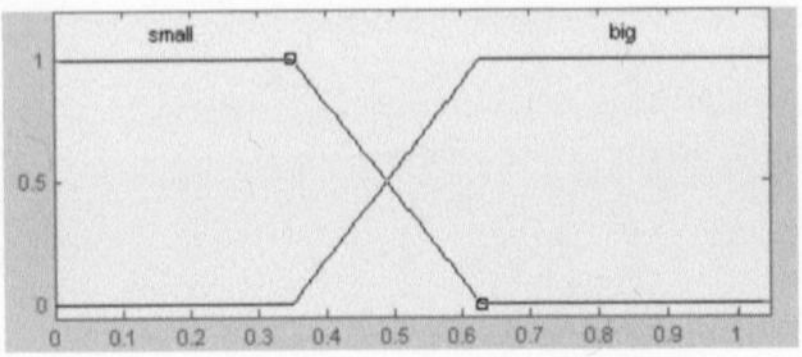

Fig. 4.21. Membership function of $\lvert\alpha_{\mathrm{G}}\rvert$

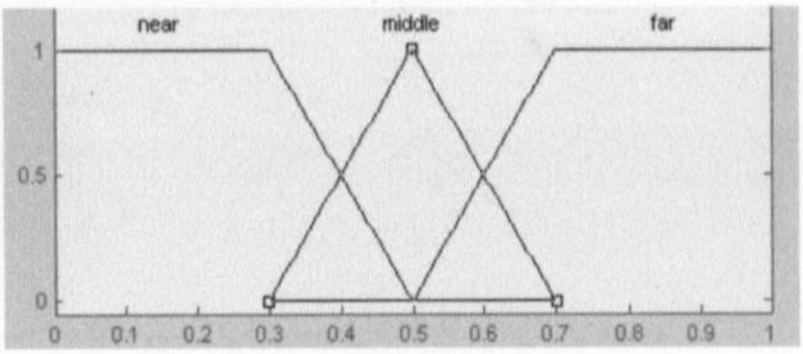

Fig. 4.22. Membership function of $d_{\min}$

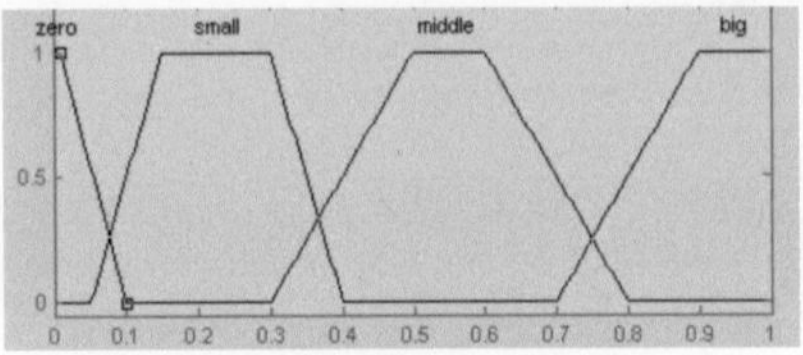

Fig. 4.23. Membership function of k_{MO}

The output of the fuzzy-based kinematic fusion module is the velocity defined as

$$\mathbf{v} = [v, \omega], \tag{4.27}$$

where

- $v = k_{\mathrm{MO}} \cdot v_{\mathrm{MO}} + (1 - k_{\mathrm{MO}}) \cdot v_{\mathrm{PP}}$,
- $\omega = k_{\mathrm{MO}} \cdot \omega_{\mathrm{MO}} + (1 - k_{\mathrm{MO}}) \cdot \omega_{\mathrm{PP}}$,
- v_{MO} and v_{PP} are the linear velocities of the collision-free behavior of the MOFIS module and the goal-oriented behavior of the PAPFIS module, respectively,
- ω_{MO} and ω_{PP} are the angular velocities of the collision-free behavior of the MOFIS module and the goal-oriented behavior of the PAPFIS module, respectively.

- k_{MO} is a behavioral weight, i.e. multiplier factor, computed from the absolute value of a goal-oriented angle $|\alpha_G|$ and a minimum obstacle distance d_{min}, $d_{min} = \min(\mathbf{d}_O)$, by the fuzzy rules of the module MOFIS-03, as listed in Fig. 4.20. The membership functions of $|\alpha_G|$, d_{min} and k_{MO} are depicted in Fig. 4.21, 4.22, and 4.23, respectively.

The output of the fuzzy-based kinematic fusion module is a set-point $\mathbf{v} = [v, \omega]$, including linear velocity v and angular velocity ω, provided for the locomotion of the AMR.

Chapter 5
Fuzzy-based path planning (PAPFIS)

5.1. Overview

Path planning is a fundamental problem in mobile robotics. The objective of path planning is to find a collision free path that leads a robot from a start to a target in the working environment containing static and/or dynamic obstacles.

In mobile robot systems, path planning may be classified as local (sensor-based) and global (model-based) path planning.

On one hand, global path planning aims to deal with the issues of long-term planning concerned with finding the best path to the target based on foreknown information of the working environment. In other words, "Global path planning requires the environment to be completely known and the terrain should be static" (Sedigh et al. 2004).

On the other hand, local path planning has to solve problems of short-term planning involved in finding local goals and avoiding unknown obstacles. This means "local path planning is done while the robot is moving" (Sedigh et al. 2004). In other words, in local path planning, a given path may be varied because the working environment may change.

Much research on path planning has been done on dynamic environments (Mechlih, 1993), (Fraichard, 1993), (Jia and Wang, 2009) and (Tang et al. 2010). However, they use information of coordinates (Mechlih, 1993) and (Fraichard, 1993), and/or of time with "the time history of a continuous sequence of configurations between the current configuration of the robot and its goal configuration" (Jia and Wang, 2009).

In this research, the mobile robot realizes a topological path planning concerning both local and global levels without metric information of coordinates. The key idea of this topological path planning is the coordination of global path planning based on a topological map that requires no metric information about the coordinates of the robot and the dimensions of the working environment, and local behavior-based path planning with the fuzzy-based behaviors for the goal-oriented navigation of mobile robots.

Fig. 5.1 illustrates the organization of the PAPFIS module. The PAPFIS module composes of three sub-modules: the PAPFIS path planner, the PAPFIS behavior and the PAPFIS behavior selector.

The path planner contains two sub-modules: global planner and local planner.

In the global planner, based on the adjacent matrix of a topological map **Ad**, the path search module computes an available path *pth* from a start point *startP* to an end point *endP* given by the user with respect to the objective of a navigation task. After that, the goal definition module defines a local goal *lGoal* controlled by a setting goal command set_G of the local planner.

In the local planner, the goal verification module verifies a found-goal state $state_G$ based on the perceptional information S_L from the OFFIS module, the inertial information of robot direction θ_R measured by auxiliary sensors (Aux.), e.g. gyroscope, and an odometry distance calculated from the linear velocity v of the set-point, i.e. the output of the MOFIS module, and the node attributes Σ_N of the topological map. Concurrently, the guidance action module identifies a goal-oriented angle α_G from a current robot direction to a goal direction, and a motion action act_G at a local goal. If the local goal is found and achieved, the goal update module will send a setting goal command set_G to enable the global planner to update a new goal.

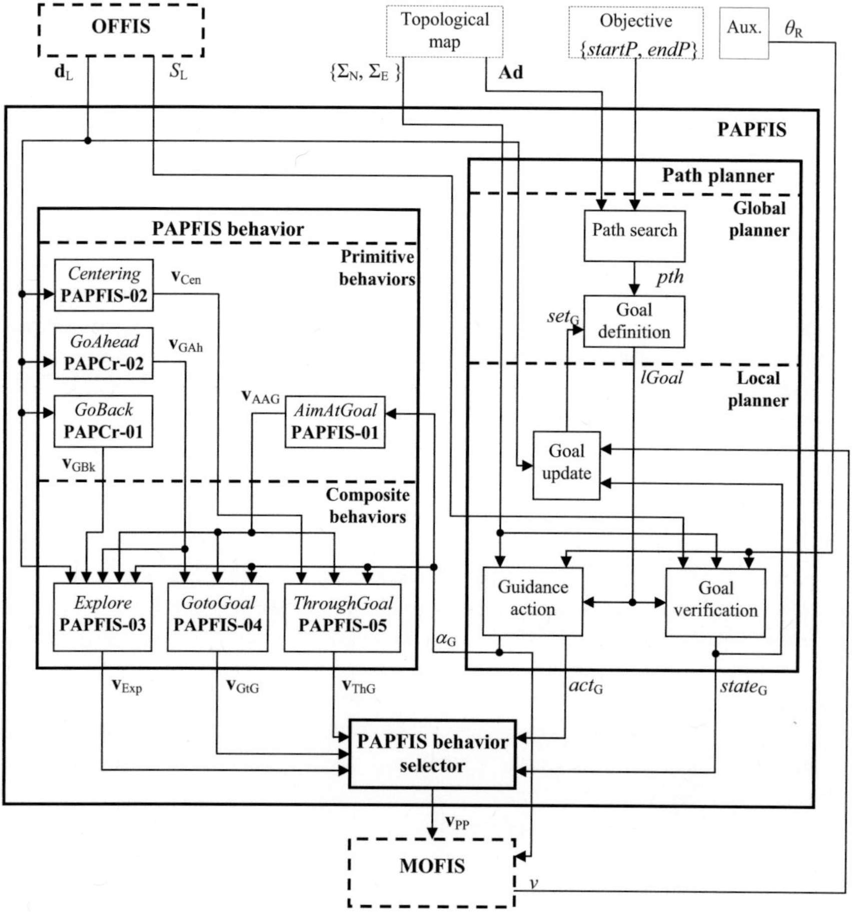

Fig. 5.1. Organization of the PAPFIS module

Based on a fuzzy landmark distance $\mathbf{d}_L$ and a goal-oriented angle α_G defined by the path planner, the PAPFIS behavior module computes the primitive and composite behaviors related to topological path planning.

The PAPFIS behavior selector uses a goal-found state $state_G$ and a robot action motion act_G at the desired goal as the decision-making information to perform automatically and autonomously a competitive coordination among the PAPFIS behaviors to activate a behavior suitable for the situation at hand. The hierarchical architecture of the PAPFIS behaviors is illustrated in Fig. 5.2. The PAPFIS behaviors contain primitive behaviors, including *GoBack*, *GoAhead*, *AimAtGoal* and *Centering*, and composite behaviors, including *Explore*, *GotoGoal* and *ThroughGoal*.

In the case that the perception (OFFIS) does not find a landmark, the $state_G$ is set to zero and the behavior *Explore* will be activated. This helps a mobile robot explore the working environment to search the desired goal. In the existing prototype implementation the robot will stop and raise an exception message, if the behavior *Explore* does not find the desired goal within a user-given time frame. There could be several strategies implemented, to remove the robot from such a deadlock situation, e.g. going back to the previous successfully found landmark and retrying the exploration.

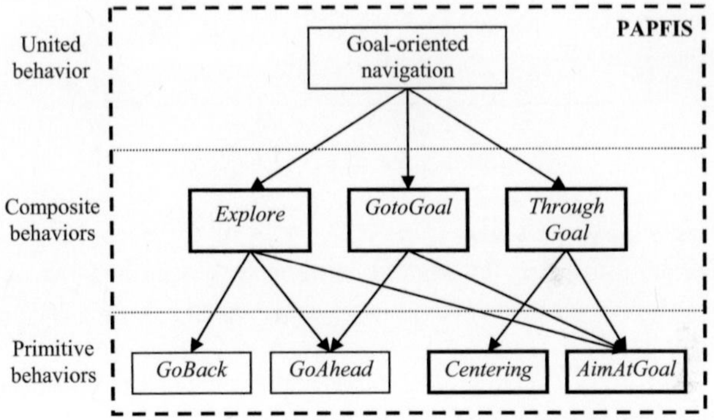

Fig. 5.2. Hierarchical architecture of the PAPFIS behaviors

The outputs of the PAPFIS module are the velocity command $\mathbf{v}_{PP}$ for a goal-oriented behavior and the goal-oriented angle α_G provided for the fuzzy-based fusion in the MOFIS module.

In the following sections, the concept of topological map will be initially presented. After that, the operations of the global planner and local planner will be introduced. Subsequently, the primitive and composite behaviors of PAPFIS are described. Finally, the PAPFIS behavior selector will be depicted.

5.2. Concept of topological map

5.2.1. Definition of variables

Let

- *LmkF* denote a landmark feature,

- θ_G denote a goal direction, where $\theta_G \in$ {N, E, S, W} are the four basic directions of the compass frame {C}, shown in Fig. 5.3. It is possible to define θ_G in an angular relationship with respect to the X_I-axis of the inertial frame {I} by

$$\theta_G = \begin{cases} \text{N} & \text{if } \tilde{\theta}_G \in [-45°, 45°] \\ \text{E} & \text{if } \tilde{\theta}_G \in (-45°, -135°) \\ \text{S} & \text{if } \tilde{\theta}_G \in [-135°, -180°] \vee [135°, 180°] \\ \text{W} & \text{if } \tilde{\theta}_G \in (45°, 135°) \end{cases}. \tag{5.1}$$

- Σ_N denote a set of node attributes, $\Sigma_N = \{N_i, LmkF_i\}$, where N_i is a node ID, $i \in \mathbb{N}$,
- Σ_E denote a set of edge attributes, $\Sigma_E = \{E_{ij}, ActF_{ij}\}$, where E_{ij} is an edge ID between nodes i and j, $i \neq j$, $ActF_{ij}$ denotes an action feature for moving from N_i to N_j.

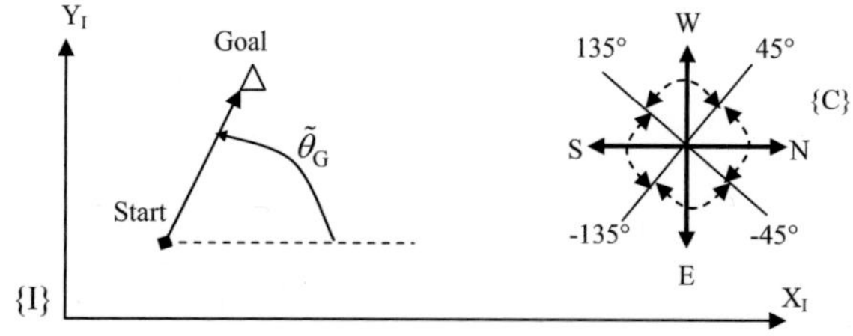

Fig. 5.3. Inertial frame {I} and compass frame {C}

5.2.2. Definition of topological map

In mobile robotics, mapping the working environments is an important task for path planning. For indoor map generation, there are two major paradigms for mapping: grid-based and graph-based maps.

According to (Florczyk, 2005), grid-based maps use a coordinate system, e.g. polar or Cartesian coordinate system, to represent the environment. Grid-based maps contain plenty of cells, small rectangular grids occupied by a value ranging from 0 to 1. The problem of grid-based maps is that they are based on metric maps, which are a set of coordinates in Cartesian space. If the search spaces are large, the computational costs are very high.

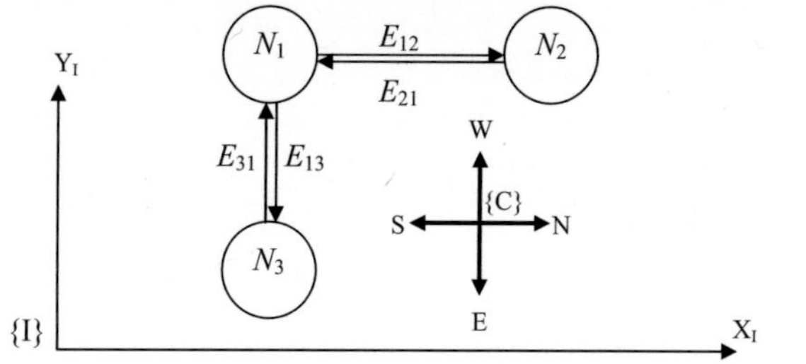

Fig. 5.4. Three-node topological map

On the contrary, graph-based maps use topological maps to solve many abstract tasks of path planning with poor metric information. The computational costs of graph-based methods are considerably smaller than that of grid-based methods.

A ***topological map*** describes a map based on a graph structure where "the graph *nodes* define landmarks such as a particular position in the environments and the graph *edges* define procedural information for moving between nodes" (Tarutoko et al. 2006). It uses an adjacency matrix **Ad**, i.e. a two-dimensional array for representing graph connections between the nodes.

Consider three-node topological map in Fig. 5.4 as an example, with the assumption that N_1 is an open door, N_2 is a corner, and N_3 is a corridor.

Assume the node and edge attributes of the topological map can be defined as shown in Fig. 5.5.

Node ID	**Landmark feature** *LmkF*	**Edge ID**	**Action feature** *ActF*
N_1	*Lmk = Door*	E_{12}	θ_G= N
		E_{13}	θ_G = E
N_2	*Lmk= Corner*	E_{21}	θ_G = S
N_3	*Lmk = Corridor*	E_{31}	θ_G= W

Fig. 5.5. Node and edge attributes of the three-node topological map

The node and edge attributes can be described by the following sets:

- $N_i \in \{N_1, N_2, N_3\}$ is the set of node ID,
- $E_{ij} \in \{E_{12}, E_{21}, E_{13}, E_{31}\}$ is the set of edge ID,
- $LmkF_i = \{Lmk\}$, where Lmk is a landmark type predefined for node N_i,
- $ActF_{ij} = \{\theta_G\}$, where θ_G is a goal direction from node N_i to node N_j.

The adjacency matrix of this topological map is formed by

$$\mathbf{Ad} = \begin{bmatrix} a_{11} & a_{12} & a_{13} \\ a_{21} & a_{22} & a_{23} \\ a_{31} & a_{32} & a_{33} \end{bmatrix} = \begin{bmatrix} 0 & 1 & 1 \\ 1 & 0 & 0 \\ 1 & 0 & 0 \end{bmatrix},$$

where, a_{ij} represents an available edge from node N_i to node N_j. If the edge exists, a_{ij} =1. Otherwise, $a_{ij} = 0$.

In general, a n-node topological map is abstractly represented by a n-by-n adjacency matrix

$$\mathbf{Ad} = \left[a_{ij} \right] = \begin{bmatrix} a_{11} & a_{12} & \cdots & a_{1n} \\ a_{21} & a_{22} & \cdots & a_{2n} \\ \vdots & \vdots & \ddots & \vdots \\ a_{n1} & a_{n2} & \cdots & a_{nn} \end{bmatrix}, \ i, j = 1 \ldots \mathrm{n}. \tag{5.2}$$

If the topological map is an undirected graph, then $\mathbf{Ad}^{\mathrm{T}} = \mathbf{Ad}$, i.e. $\mathbf{Ad}$ is symmetric. In other cases, $\mathbf{Ad}^{\mathrm{T}} \neq \mathbf{Ad}$, i.e. $\mathbf{Ad}$ is asymmetric.

5.3. Global planner

5.3.1. Graph search algorithms

In artificial intelligence, there are two major groups of graph search strategies including uninformed search strategies and informed search strategies.

According to (Russell and Norvig, 2003), the *uninformed search* strategies, also called blind search strategies, need no additional information about states beyond that provided in the problem definition; The *informed search* strategies, also called *heuristic search* strategies, use problem-specific knowledge beyond the definition of the problem itself.

Basically, the informed search strategies find solutions more efficiently than the uninformed search strategies.

Recently, A* search algorithms, belonging to informed search strategies, have been widely developed for path planning in robotics (Fraichard, 1993), (Chiang et al. 2007). Although the A* algorithms are really effective, complete and optimal, they strongly depend on metric information. That means the A* search is not suitable for topological path planning without metric information.

In contrast to the A* algorithms, iterative deepening search strategies, members of uninformed search strategies, need no metric information. For this reason, iterative deepening search strategies can help behavior-based control with topological path planning in "being more simple and natural" (Jia and Wang, 2009).

The iterative deepening search is the development of depth-first search, where finding the best *depth limit* is found by increasing the depth limit gradually until a goal is discovered. In other words, iterative deepening search combines "the benefits of the depth-first search (the modest memory) and breadth-first search (complete and optimal)" (Russell and Norvig, 2003).

As affirmed in (Russell and Norvig, 2003), the iterative deepening search is the preferred uninformed search method when there is a large search space because it is possibly complete and optimal without searching the entire graph space.

More details of the iterative deepening search can be found in (Russell and Norvig, 2003).

5.3.2. Path search

Based on the adjacency matrix **Ad**, the path search module carries out a graph search algorithm (Russell and Norvig, 2003) to find a global path from a start node (point) *startP* to an end node *endP* given by the user in the objective module, see Fig. 5.1. The global path is formed

$$pth = \{[N_i, N_j] | a_{ij} = 1, (i \neq j) \in \mathbb{N}\}, \tag{5.3}$$

where $pth(1) = startP$ and $pth(\text{end}) = endP$.

For example, the global path from N_2 to N_3 in the topological map illustrated in Fig. 3 is $pth = [N_2, N_1, N_3]$, with $pth(1) = N_2$ and $pth(\text{end}) = N_3$.

5.3.3. Goal definition

Assume that at any given moment, a mobile robot is on its way to a derived local goal N_k. The local goal is defined by

$$lGoal = \{ pth(l) \mid pth(l) = N_k, 1 \le l \le \text{length}(pth) \}. \quad (5.4)$$

5.4. Local planner

5.4.1. Guidance action

With the assumption that an action feature *ActF* of each edge contains not only the goal direction θ_G, but also a motion action act_G that guides the robot motion at a local goal, e.g. going to the goal or going through the goal, i.e. $ActF_{ij} = \{\theta_G, act_G\}$, where θ_G is a goal direction from node N_i to node N_j, and act_G is a motion action at node N_j.

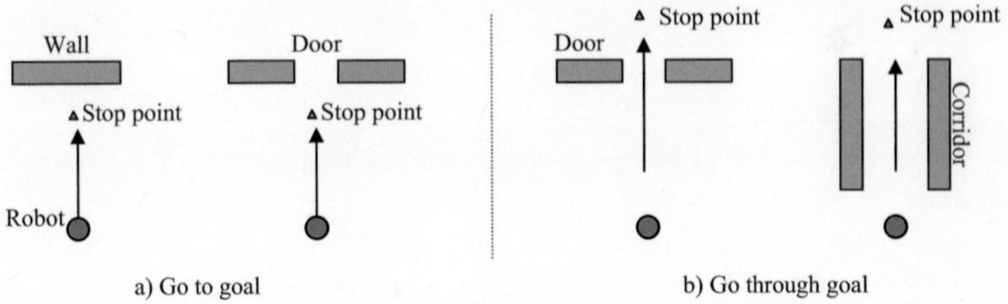

Fig. 5.6. Robot motion actions at different goals
a) $act_G = Goto$; b) $act_G = Through$

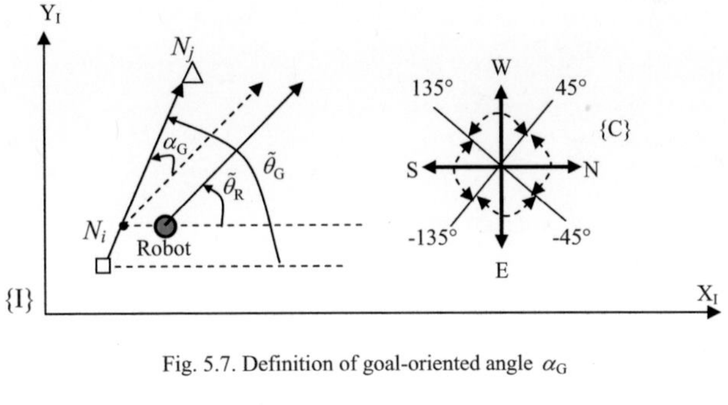

Fig. 5.7. Definition of goal-oriented angle α_G

A guidance action is used to navigate a mobile robot to a local goal from any current position of the robot. The guidance action module computes a guidance action $GdAct$, including a goal-oriented angle α_G and a motion action act_G at a local goal, based on the action feature *ActF* as follows:

$$GdAct = \{\alpha_G, act_G\}, \quad (5.5)$$

where

- $act_{\mathrm{G}} = \begin{cases} Goto & \text{for going to goal} \\ Through & \text{for going through goal} \end{cases}$, illustrated in Fig. 5.6,
- α_{G} is the goal-oriented angle, $\alpha_{\mathrm{G}} = \theta_{\mathrm{G}} - \theta_{\mathrm{R}}$,
- θ_{R} is the robot direction, where $\theta_{\mathrm{R}} \in$ {N, E, S, W} are the four basic directions of the compass frame {C}, shown in Fig. 5.7. It is possible to define θ_{R} in an angular relationship with respect to the X_{I}-axis of the inertial frame {I} by

$$\theta_{\mathrm{R}} = \begin{cases} \mathrm{N} & \text{if } \tilde{\theta}_{\mathrm{R}} \in [-45°, 45°] \\ \mathrm{E} & \text{if } \tilde{\theta}_{\mathrm{R}} \in (-45°, -135°) \\ \mathrm{S} & \text{if } \tilde{\theta}_{\mathrm{R}} \in [-135°, -180°] \vee [135°, 180°] \\ \mathrm{W} & \text{if } \tilde{\theta}_{\mathrm{R}} \in (45°, 135°) \end{cases},$$

- θ_{G} is the goal direction, which is also defined through the compass frame {C} by

$$\theta_{\mathrm{G}} = \begin{cases} \mathrm{N} & \text{if } \tilde{\theta}_{\mathrm{G}} \in [-45°, 45°] \\ \mathrm{E} & \text{if } \tilde{\theta}_{\mathrm{G}} \in (-45°, -135°) \\ \mathrm{S} & \text{if } \tilde{\theta}_{\mathrm{G}} \in [-135°, -180°] \vee [135°, 180°] \\ \mathrm{W} & \text{if } \tilde{\theta}_{\mathrm{G}} \in (45°, 135°) \end{cases}.$$

The information *GdAct* is supported with the computation of the PAPFIS and MOFIS behaviors.

5.4.2. Goal verification

A goal is defined to be found if the two following conditions are satisfied:

- a recognized landmark type of observed object is the same as the expected landmark type predefined for the goal,
- a robot direction θ_{R} approximates a goal direction θ_{G}, i.e. $\theta_{\mathrm{G}} \approx \theta_{\mathrm{R}}$.

A goal-found state $state_{\mathrm{G}}$ is set to one if a goal is found, and set to zero if no goal found. In accordance with the definition, the goal-found state $state_{\mathrm{G}}$ is defined as

$$state_{\mathrm{G}} = \begin{cases} 1 & \text{if } (S_{\mathrm{L}} \equiv Lmk) \wedge (|\alpha_{\mathrm{G}} = \theta_{\mathrm{G}} - \theta_{\mathrm{R}}| < \varepsilon) \\ 0 & \text{otherwise} \end{cases}, \tag{5.6}$$

where

- S_{L} is the recognized landmark type,
- Lmk is the expected landmark type,
- ε is an user-given offset.

5.4.3. Goal update

A goal is defined to be achieved if one of the two following conditions is satisfied:

- a distance from the robot to an impassable goal, e.g. walls and corners, is smaller than a safe distance given by the user,
- a movement distance when the robot goes through a passable goal, e.g. open doors and corridors, is bigger than a goal depth given by the user.

After a local goal is found and achieved, the goal update module generates a setting goal command set_{G} to enable the goal definition module to update a new goal.

For the impassable goals with $act_G = Goto$, the setting goal command is defined by

$$set_G = \begin{cases} 1 \text{ if } (state_G = 1) \wedge (\min(\mathbf{d}_L) < \delta_G) \\ 0 \text{ otherwise} \end{cases}, \tag{5.7}$$

where

- $\mathbf{d}_L = [d_{L1} \quad d_{L2} \quad d_{L3} \quad d_{L4}]$ is a landmark distance defined in the equation (3.49) in chapter 3,
- δ_G is the safe distance to a goal given by the user.

For the passable goals with $act_G = Through$, the setting goal command is formed as

$$set_G = \begin{cases} 1 \text{ if } (state_G = 1) \wedge (\rho_R > Dp_G) \\ 0 \text{ otherwise} \end{cases}, \tag{5.8}$$

where

- ρ_R is the movement distance calculated from the linear velocity v of the set-point,
- Dp_G is the goal depth, which is approximately defined by the user.

The setting goal command set_G plays the role as a decision signal to define a new local goal in the global planner as the follows

$$lGoal = \{pth(l + set_G) | 1 \leq l + set_G \leq \text{length}(pth)\}, \tag{5.9}$$

where l is the index of the global path *pth*, and *pth*(l) supports the node ID of a current local goal that the robot is on its way to.

5.5. PAPFIS primitive behaviors

The primitive PAPFIS behaviors comprise *GoBack*, *GoAhead*, *AimAtGoal* and *Centering*.

The two first primitive behaviors are monotonic because their motion direction is unchanged and independent of the stimulus strength. These behaviors use crisp data, so they are symbolized with PAPCr-01 and PAPCr-02, as illustrated in Fig. 5.1.

The velocity commands of the primitive behaviors *GoBack* and *GoAhead* have respectively the forms

$$\mathbf{v}_{GBk} = [v_{GBk}, \omega_{GBk}], \tag{5.10}$$

$$\mathbf{v}_{GAh} = [v_{GAh}, \omega_{GAh}], \tag{5.11}$$

where

- v_{GBk} and v_{GAh} are the linear velocities, $v_{GBk} = -1 + \frac{d_{Ls}}{2}$, $v_{GAh} = d_{Ls}$, $d_{Ls} = \min(\mathbf{d}_L)$,
- ω_{GBk} and ω_{GAh} are the angular velocities, $\omega_{GBk} = 0$, $\omega_{GAh} = 0$.

The other two primitive behaviors *AimAtGoal* and *Centering* are fuzzy-based behaviors because they use a fuzzy inference system to calculate the kinematics.

The primitive behavior *AimAtGoal* is responsible for tuning a robot direction to a goal direction. The velocity command of the primitive behavior *AimAtGoal* has the form

$$\mathbf{v}_{AAG} = [v_{AAG}, \omega_{AAG}], \tag{5.12}$$

where $v_{AAG} = 0$ and ω_{AAG} is calculated from the goal-oriented angle α_G by the fuzzy rules of the PAPFIS-01, shown in Fig. 5.8. The membership functions of α_G and ω_{AAG} are respectively illustrated in Fig. 5.9 and Fig. 5.10.

The primitive behavior *Centering* controls the balancing movement of the robot in hollow paths, imitating the balancing behavior of bees for flight in a corridor. Consider the situation in Fig. 5.11 as an example of the primitive behavior *Centering*.

Fuzzy rules of PAPFIS-01

IF	***THEN***
α_G is bigNeg	ω_{AAG} is fastLeft
α_G is midNeg	ω_{AAG} is fastLeft
α_G is smallNeg	ω_{AAG} is smallLeft
α_G is zero	ω_{AAG} is zero
α_G is smallPos	ω_{AAG} is smallRight
α_G is midPos	ω_{AAG} is fastRight
α_G is bigPos	ω_{AAG} is fastRight

Fig. 5.8. Fuzzy rules of PAPFIS-01

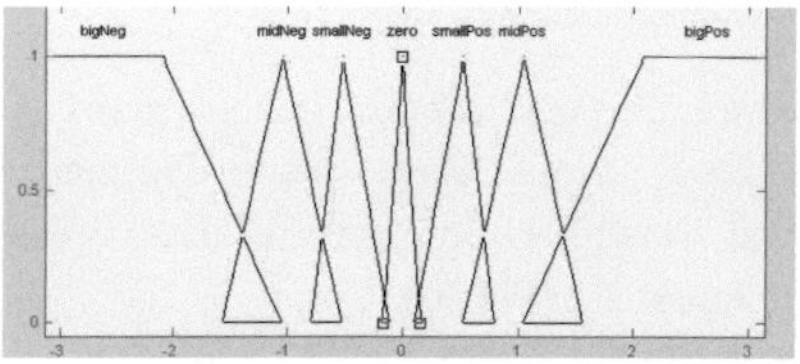

Fig. 5.9. Membership function of α_G

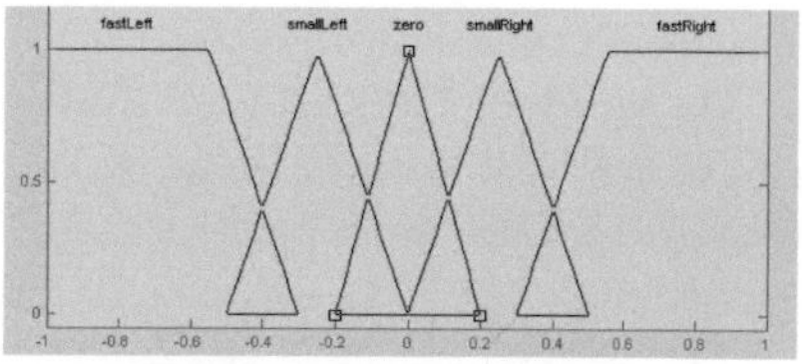

Fig. 5.10. Membership function of ω_{AAG}

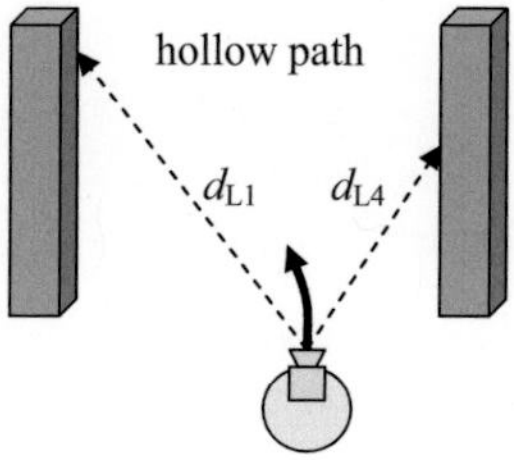

Fig. 5.11. Balancing movement in a hollow path

The module PAPFIS-02 of the primitive behavior *Centering* is illustrated in Fig. 5.12.

From the landmark distance $\mathbf{d}_L$, the parameter calculation module calculates two deduced parameters as follows

$$k_d = \mathbf{d}_L(1) / \mathbf{d}_L(\text{end}) = d_{L1} / d_{L4}, \tag{5.13}$$

$$d_{min} = \min\left(\mathbf{d}_L(1), \mathbf{d}_L(\text{end})\right) = \min\left(d_{L1}, d_{L4}\right). \tag{5.14}$$

The velocity of the primitive behavior *Centering* has the form

$$\mathbf{v}_{Cen} = \left[v_{Cen}, \omega_{Cen}\right], \tag{5.15}$$

where v_{Cen} and ω_{Cen} are defined from the two deduced parameters, k_d and d_{min}, by the fuzzy logic rules of the PAPFIS-02, shown in Fig. 5.13.

The membership functions of k_d and d_{min} are shown respectively in Fig. 5.14 and 5.15. The membership functions of v_{Cen} and ω_{Cen} are displayed respectively in Fig. 5.16 and 5.17.

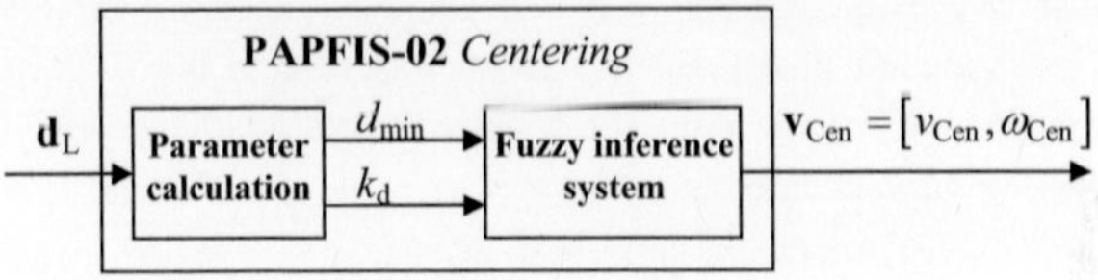

Fig. 5.12. Fuzzy-based module of the primitive behavior *Centering*

Fuzzy rules of PAPFIS-02

IF	*THEN*
d_{min} is very near	v_{Cen} is slow
d_{min} is near	v_{Cen} is slow
d_{min} is middle	v_{Cen} is medium
d_{min} is far	v_{Cen} is fast
d_{min} is very-far	v_{Cen} is fast
k_d is very smaller	ω_{Cen} is fastLeft
k_d is smaller	ω_{Cen} is smallLeft
k_d is equal	ω_{Cen} is zero
k_d is bigger	ω_{Cen} is smallRight
k_d is very bigger	ω_{Cen} is fastRight

Fig. 5.13. Fuzzy rules of PAPFIS-02

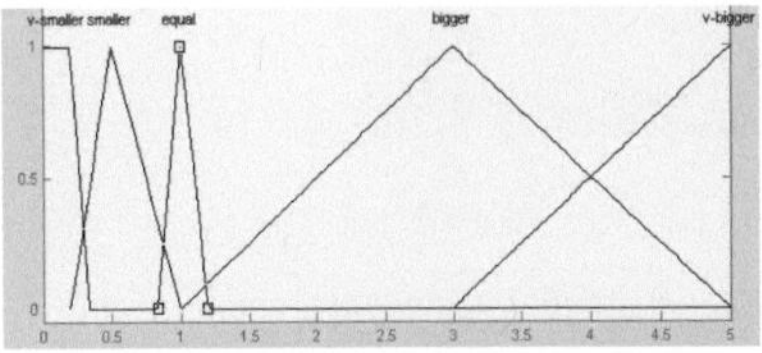

Fig. 5.14. Membership function of k_d

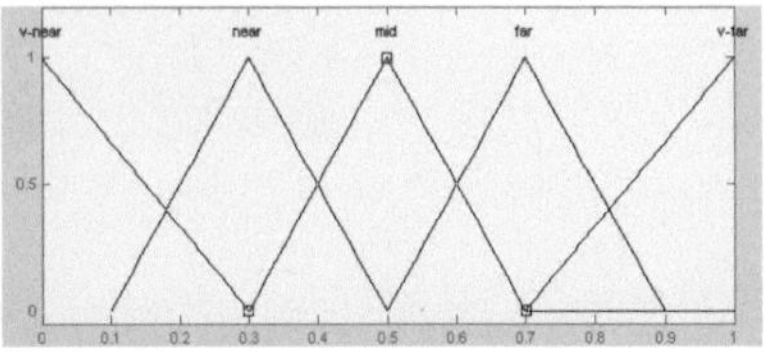

Fig. 5.15. Membership function of d_{min}

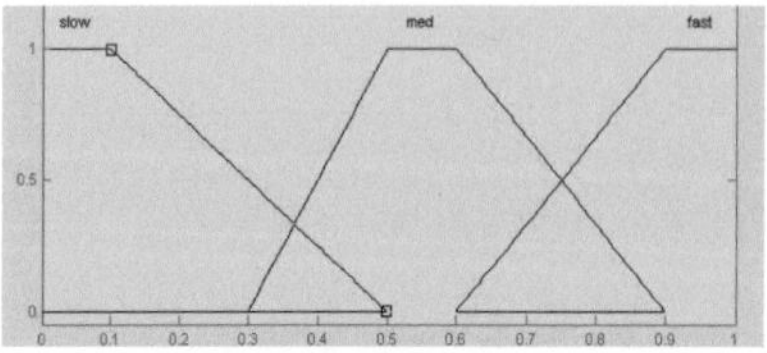

Fig. 5.16. Membership function of v_{Cen}

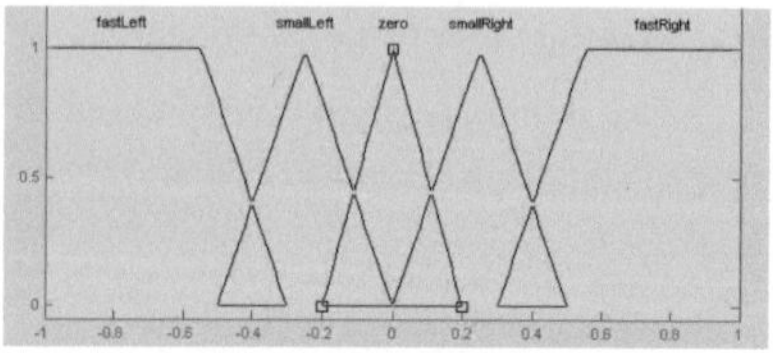

Fig. 5.17. Membership function of ω_{Cen}

5.6. PAPFIS composite behaviors

The composite behaviors of the PAPFIS include *Explore*, *GotoGoal* and *ThroughGoal*.

The composite behavior *Explore* drives a mobile robot to explore the working environment to find a specific goal based on a given compass direction of the goal.

The module PAPFIS-03 of the composite behavior *Explore* is illustrated in Fig. 5.18 with five inputs, including three competitive velocities $\mathbf{v}_{\mathrm{GBk}}$, $\mathbf{v}_{\mathrm{GAh}}$, and $\mathbf{v}_{\mathrm{AAG}}$, and two measured parameters $\mathbf{d}_{\mathrm{L}}$ and α_{G} to compute the sum coefficients.

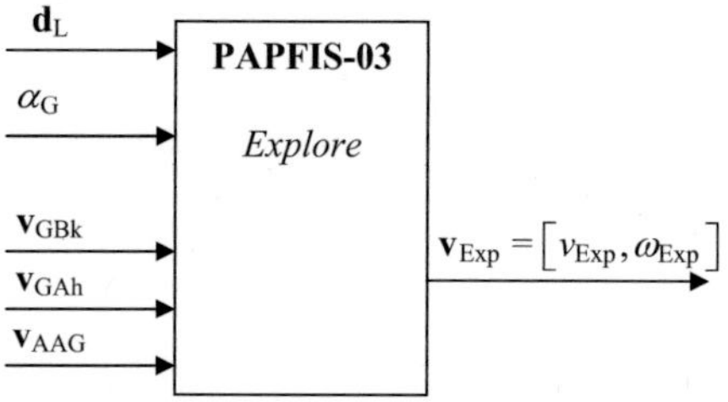

Fig. 5.18. Fuzzy-based module of the composite behavior *Explore*

Fuzzy rules of PAPFIS-03

IF	***THEN***
$\lvert\alpha_{\mathrm{G}}\rvert$ is zero	k_{E1} is zero
$\lvert\alpha_{\mathrm{G}}\rvert$ is small	k_{E1} is small
$\lvert\alpha_{\mathrm{G}}\rvert$ is medium	k_{E1} is middle
$\lvert\alpha_{\mathrm{G}}\rvert$ is large	k_{E1} is big
$\lvert\alpha_{\mathrm{G}}\rvert$ is very large	k_{E1} is very big
$d_{\max}$ is close	k_{E2} is zero
$d_{\max}$ is not close	k_{E2} is one

Fig. 5.19. Fuzzy rules of PAPFIS-03

The velocity command of the composite behavior *Explore* has the form

$$\mathbf{v}_{\mathrm{Exp}} = \left[v_{\mathrm{Exp}}, \omega_{\mathrm{Exp}}\right], \tag{5.16}$$

where:

- $v_{\text{Exp}} = k_{\text{E1}} \cdot v_{\text{AAG}} + (1 - k_{\text{E1}}) \cdot (k_{\text{E2}} \cdot v_{\text{GAh}} + (1 - k_{\text{E2}}) \cdot v_{\text{GBk}})$, (5.17)
- $\omega_{\text{Exp}} = k_{\text{E1}} \cdot \omega_{\text{AAG}} + (1 - k_{\text{E1}}) \cdot (k_{\text{E2}} \cdot \omega_{\text{GAh}} + (1 - k_{\text{E2}}) \cdot \omega_{\text{GBk}})$, (5.18)
- k_{E1} and k_{E2} are the multiplier factors determined from $|\alpha_{\text{G}}|$ and d_{max} by the fuzzy rules listed in Fig. 5.19, the membership functions of k_{E1} and k_{E2} are respectively shown in Fig. 5.22 and Fig. 5.23,
- $|\alpha_{\text{G}}|$ is an absolute value of a goal-oriented angle α_{G}, its membership function is shown in Fig. 5.20,
- $d_{\text{max}} = \max(\mathbf{d}_{\text{L}})$, its membership function is shown in Fig. 5.21.

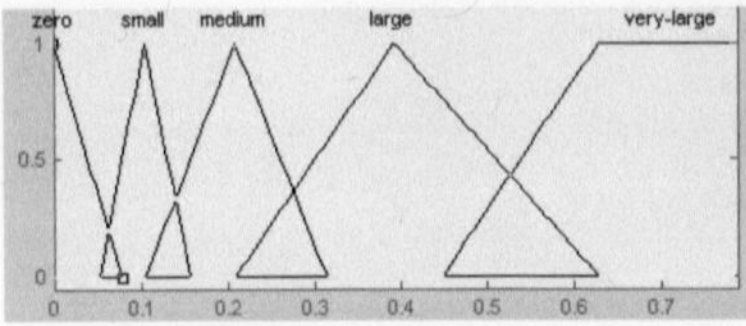

Fig. 5.20. Membership function of input $|\alpha_{\text{G}}|$

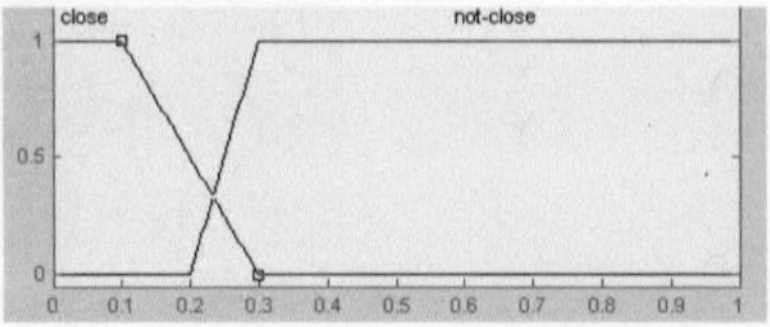

Fig. 5.21. Membership function of input d_{max}

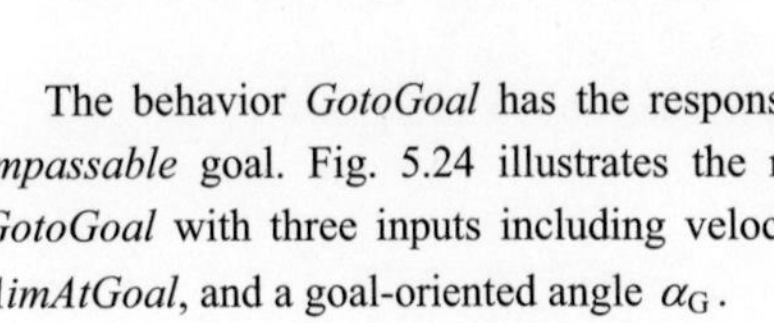

Fig. 5.22. Membership function of output k_{E1}

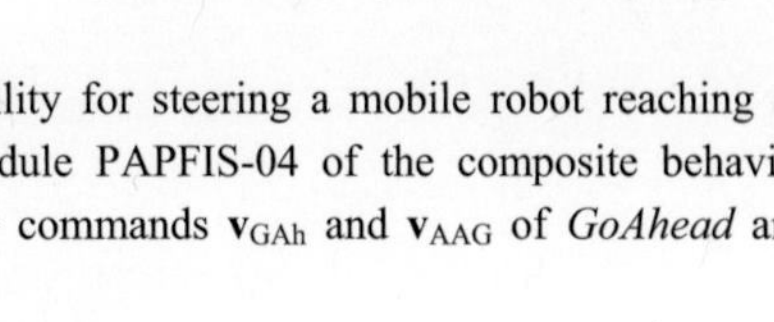

Fig. 5.23. Membership function of output k_{E2}

The behavior *GotoGoal* has the responsibility for steering a mobile robot reaching an *impassable* goal. Fig. 5.24 illustrates the module PAPFIS-04 of the composite behavior *GotoGoal* with three inputs including velocity commands $\mathbf{v}_{\text{GAh}}$ and $\mathbf{v}_{\text{AAG}}$ of *GoAhead* and *AimAtGoal*, and a goal-oriented angle α_{G}.

The velocity of the behavior *GotoGoal* has the form

$$\mathbf{v}_{\text{GtG}} = [v_{\text{GtG}}, \omega_{\text{GtG}}], \quad (5.19)$$

where:

- $v_{\text{GtG}} = k_{\text{GtG}} \cdot v_{\text{AAG}} + (1 - k_{\text{GtG}}) \cdot v_{\text{GAh}}$, (5.20)
- $\omega_{\text{GtG}} = k_{\text{GtG}} \cdot \omega_{\text{AAG}} + (1 - k_{\text{GtG}}) \cdot \omega_{\text{GAh}}$, (5.21)
- k_{GtG} is a multiplier factor derived from absolute value of a goal-oriented angle α_{G} by the fuzzy rules listed in Fig. 5.25. The membership function of k_{GtG} is the same as that of k_{E1}, shown in Fig. 5.22, and that of $|\alpha_{\text{G}}|$ is shown in Fig. 5.20.

Fig. 5.24. Fuzzy-based module of the composite behavior *GotoGoal*

Fuzzy rules of PAPFIS-04

IF	*THEN*
$\|\alpha_G\|$ is zero	k_{GtG} is zero
$\|\alpha_G\|$ is small	k_{GtG} is small
$\|\alpha_G\|$ is medium	k_{GtG} is middle
$\|\alpha_G\|$ is large	k_{GtG} is big
$\|\alpha_G\|$ is very large	k_{GtG} is very big

Fig. 5.25. Fuzzy rules of PAPFIS-04

Unlike the behavior *GotoGoal*, the behavior *ThroughGoal* is responsible for guiding a mobile robot through a hollow path considered to be a *passable* goal, e.g. open doors or corridors. For this reason, it is necessary to invoke the primitive behavior *Centering* to the composite behavior *ThroughGoal*.

Fig. 5.26. Fuzzy-based module of the composite behavior *ThroughGoal*

Fuzzy rules of PAPFIS-05

IF	*THEN*
$\|\alpha_G\|$ is zero	k_{ThG} is zero
$\|\alpha_G\|$ is small	k_{ThG} is zero
$\|\alpha_G\|$ is medium	k_{ThG} is small
$\|\alpha_G\|$ is large	k_{ThG} is small
$\|\alpha_G\|$ is very large	k_{ThG} is very big

Fig. 5.27. Fuzzy rules of PAPFIS-05

The module PAPFIS-05 of the composite behavior *ThroughGoal* illustrated in Fig. 5.26 has three inputs containing the velocity $\mathbf{v}_{\mathrm{Cen}}$ of *Centering*, velocity $\mathbf{v}_{\mathrm{AAG}}$ of *AimAtGoal*, and an absolute value of the different angle α_{G}.

The velocity of *ThroughGoal* has the form

$$\mathbf{v}_{\mathrm{ThG}} = \left[v_{\mathrm{ThG}}, \omega_{\mathrm{ThG}} \right], \quad (5.22)$$

where:

- $v_{\mathrm{ThG}} = k_{\mathrm{ThG}} \cdot v_{\mathrm{AAG}} + \left(1 - k_{\mathrm{ThG}}\right) \cdot v_{\mathrm{Cen}}$, (5.23)
- $\omega_{\mathrm{ThG}} = k_{\mathrm{ThG}} \cdot \omega_{\mathrm{AAG}} + \left(1 - k_{\mathrm{ThG}}\right) \cdot \omega_{\mathrm{Cen}}$, (5.24)
- k_{ThG} is a multiplier factor for weighting the primitive behaviors of *ThroughGoal*, which is defined from $|\alpha_{\mathrm{G}}|$ by the fuzzy inference system in the PAPFIS-05, illustrated in Fig. 5.27. The membership function of k_{ThG} is the same as that of k_{E1} shown in Fig. 5.22.

5.7. PAPFIS behavior selector

It is obvious that the behaviors *Explore*, *GotoGoal*, and *ThoughGoal* have the different and even contradicting mission objectives. For this reason, only one of them is activated exclusively for a certain situation.

The behavior selector of PAPFIS plays the role as an arbitrator to resolve the conflict arising from these behaviors. It uses $state_{\mathrm{G}}$ and act_{G} of the local planner as decision-making information to automatically and autonomously perform a competitive coordination among the three composite behaviors *Explore*, *GotoGoal*, and *ThoughGoal*.

The PAPFIS behavior selector module is illustrated in Fig. 5.28 with three data input signals including the velocity commands $\mathbf{v}_{\mathrm{Exp}}$, $\mathbf{v}_{\mathrm{GtG}}$, and $\mathbf{v}_{\mathrm{ThG}}$, and two decision signals, including a goal-found state $state_{\mathrm{G}}$ and a motion action act_{G}.

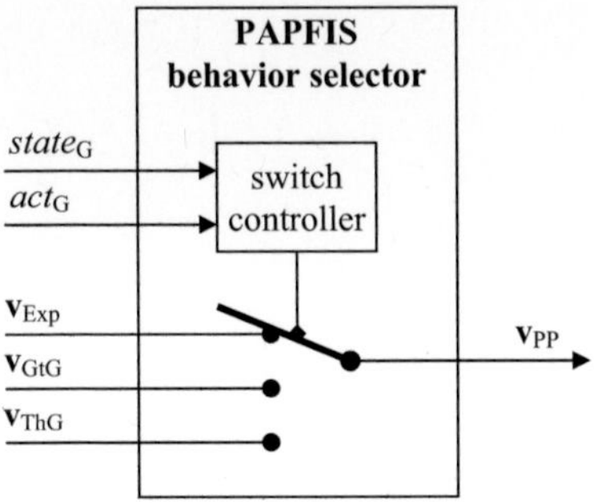

Fig. 5.28. PAPFIS behavior selector

The output of the selector is the velocity of the goal-oriented behavior defined by

$$\mathbf{v}_{\mathrm{PP}} = \left[v_{\mathrm{PP}}, \omega_{\mathrm{PP}} \right], \quad (5.25)$$

where:

$$- \; v_{\mathrm{PP}} = \begin{cases} v_{\mathrm{Exp}} & \text{if } \left(state_{\mathrm{G}} = 0\right) \\ v_{\mathrm{GtG}} & \text{if } \left(state_{\mathrm{G}} = 1\right) \wedge \left(act_{\mathrm{G}} = Goto\right) \\ v_{\mathrm{ThG}} & \text{if } \left(state_{\mathrm{G}} = 1\right) \wedge \left(act_{\mathrm{G}} = Through\right) \end{cases}, \tag{5.26}$$

$$- \; \omega_{\mathrm{PP}} = \begin{cases} \omega_{\mathrm{Exp}} & \text{if } \left(state_{\mathrm{G}} = 0\right) \\ \omega_{\mathrm{GtG}} & \text{if } \left(state_{\mathrm{G}} = 1\right) \wedge \left(act_{\mathrm{G}} = Goto\right) \\ \omega_{\mathrm{ThG}} & \text{if } \left(state_{\mathrm{G}} = 1\right) \wedge \left(act_{\mathrm{G}} = Through\right) \end{cases}. \tag{5.27}$$

Afterwards, the velocity command $\mathbf{v}_{\mathrm{PP}}$, which is one of two outputs of the PAPFIS module, is provided for fuzzy-based kinematic fusion in the MOFIS module.

Chapter 6
Simulation based test and verification

This chapter presents test results based on simulation models in order to verify the proposed method of optical flow-based perception and the designed structure of behavior-based controller described in the previous chapters. These simulations have been implemented for different scenarios and various operating conditions, for example, for coping with static obstacles and dynamic obstacles and for working in large and narrow areas. The simulation results are illustrated by the ground truths of the simulated mobile robot with several highlighted waypoint images.

6.1. Simulation setups

6.1.1. Simulated environment

The vision-based simulated environment was established using the rendering-program POV-Ray (version 3.6) enabling 3D environment simulation. POV-Ray is the acronym for Persistence of Vision Ray-Tracer. This is a high-quality open source tool for creating stunning three-dimensional graphics. The POV-Ray program is a powerful tool used for creating three-dimensional, photo-realistic images with a ray-tracing technique.

Ray-tracing is a rendering technique for calculating an image of a scene by simulating the way rays of light travel in the real world. "Ray-tracing programs like POV-Ray start with their simulated camera and trace rays backwards out into the scene," stated in (POV-Team).

The POV-Ray program reads information about objects and lighting in a scene stored in a text file and then generates an image of the scene from the view point of a camera described in the text file. The images can be generated in a very high quality with realistic reflections, shading, perspective and other effects.

More details about POV-Ray program are given in (POV-Team).

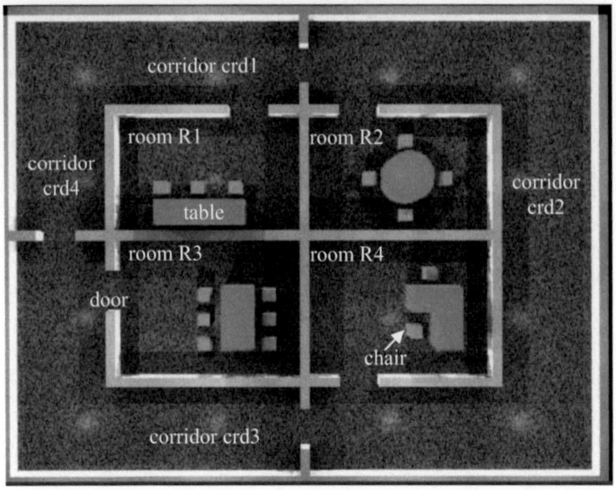

Fig. 6.1. Top view of the simulated environment in 3D with POV-Ray

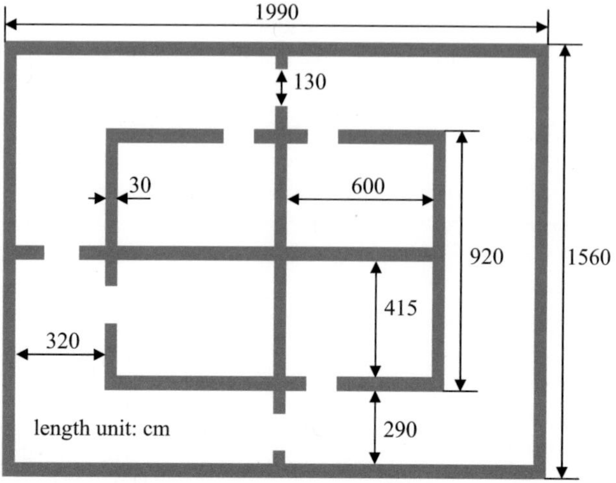

Fig. 6.2. Top view of the simulated 3D environmental model (units in centimeter)

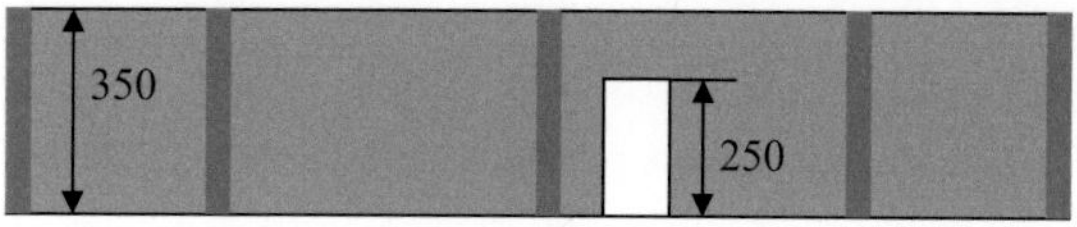

Fig. 6.3. The front view of the simulated 3D environmental model

The simulated 3D environment was constructed as an office scenario that includes four rooms, some doors, corridors, and office furniture, e.g. tables and chairs. The simulated environment is illustrated in Fig. 6.1 by the top view. The rooms were given the symbols R1, R2, R3, and R4. The corridors were given the symbols crd1, crd2, crd3, and crd4.

The dimensions of the environment with the unit of length in centimeter (cm) are shown in Fig. 6.2 and Fig. 6.3. The size of each of the room doors was 130 cm wide x 250 cm high, the size of each of the corridor doors was 130 cm wide x 350 cm high. The width of the vertical corridors, crd2 and crd4, was 320 cm, and the width of the horizon corridors, crd1 and crd3, was 290 cm.

The different dimensions of the objects in the simulated environment were aimed at testing the robustness of optical flow-based perception.

6.1.2. Topological map

The navigation tasks of the simulated experiments involved topological path planning, which required a topological map of the working environment. Notice that the topological map does not contain metric information about the dimensions and coordinates of the simulated environment. The definition of topological map includes the adjacency matrix and the attributes of nodes and edges as described in chapter 5.

Fig. 6.4 illustrates the 16-node topological map of the simulated environment. These nodes are defined by inherent indoor landmarks in the simulated office. For example, N_1, N_6, N_8, and N_{13} are corners; N_2, N_3, N_4, N_5, N_9, N_{10}, N_{11}, N_{14}, N_{15}, and N_{16} are doors; N_7 and N_{12} are corridors.

Additionally, a gyroscope was used to identify the orientation of the mobile robot. The basic directions of the gyroscope are depicted in Fig. 6.5.

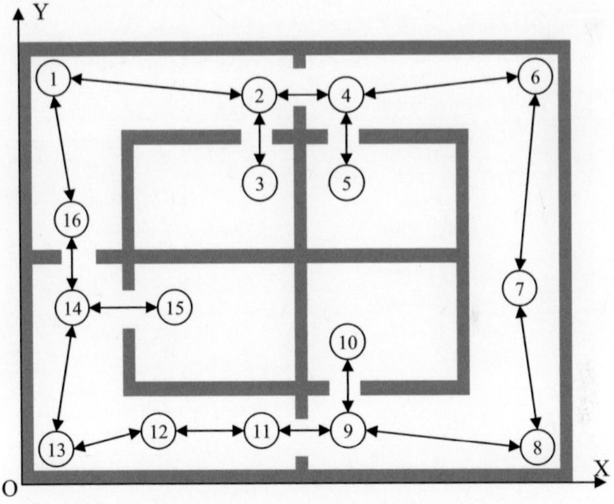

Fig. 6.4. Topological map of the simulated environment

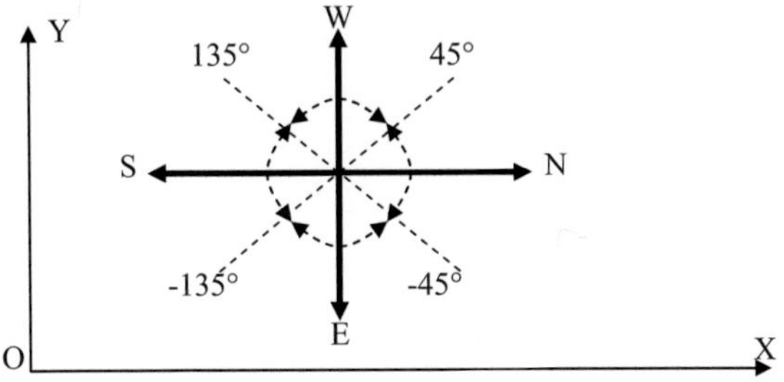

Fig. 6.5. Basic directions of gyroscope in simulations
(same coordinate system in Fig 6.4)

Node ID	Landmark feature *LmkF*	Edge ID	Action feature *ActF*
N_1	*Lmk* = *Corner* $Dp_G = 0$ cm	E_{1_2}	θ_G=N, act_G= *Goto*
		E_{1_16}	θ_G =E, act_G = *Goto*
N_2	*Lmk* = *Door* $Dp_G \approx 260$ cm	E_{2_1}	θ_G =S, act_G = *Goto*
		E_{2_3}	θ_G =E, act_G = *Through*
		E_{2_4}	θ_G =N, act_G = *Through*
N_3	*Lmk* = *Door* $Dp_G \approx 260$ cm	E_{3_2}	θ_G =W, act_G = *Through*
N_4	*Lmk* = *Door* $Dp_G \approx 260$ cm	E_{4_2}	θ_G =S, act_G = *Through*
		E_{4_5}	θ_G =E, act_G = *Through*
		E_{4_6}	θ_G =N, act_G = *Goto*
N_5	*Lmk* = *Door* $Dp_G \approx 260$ cm	E_{5_4}	θ_G =W, act_G = *Through*
N_6	*Lmk* = *Corner* $Dp_G = 0$ cm	E_{6_4}	θ_G =S, act_G = *Goto*
		E_{6_7}	θ_G =E, act_G = *Through*
N_7	*Lmk* = *Corridor* $Dp_G \approx 620$ cm	E_{7_6}	θ_G =W, act_G = *Goto*
		E_{7_8}	θ_G =E, act_G = *Goto*
N_8	*Lmk* = *Corner* $Dp_G = 0$ cm	E_{8_7}	θ_G =W, act_G = *Through*
		E_{8_9}	θ_G =S, act_G = *Goto*
N_9	*Lmk* = *Door* $Dp_G \approx 260$ cm	E_{9_8}	θ_G =N, act_G = *Goto*
		E_{9_10}	θ_G =W, act_G = *Through*
		E_{9_11}	θ_G =S, act_G = *Through*
N_{10}	*Lmk* = *Door* $Dp_G \approx 260$ cm	E_{10_9}	θ_G =E, act_G = *Through*
N_{11}	*Lmk* = *Door* $Dp_G \approx 260$ cm	E_{11_9}	θ_G =N, act_G = *Through*
		E_{11_12}	θ_G =S, act_G = *Through*
N_{12}	*Lmk* = *Corridor* $Dp_G \approx 400$ cm	E_{12_11}	θ_G =N, act_G = *Goto*
		E_{12_13}	θ_G =S, act_G = *Goto*
N_{13}	*Lmk* = *Corner* $Dp_G = 0$ cm	E_{13_12}	θ_G =N, act_G = *Through*
		E_{13_14}	θ_G =W, act_G = *Through*
N_{14}	*Lmk* = *Door* $Dp_G \approx 260$ cm	E_{14_13}	θ_G =E, act_G = *Goto*
		E_{14_15}	θ_G =N, act_G = *Through*
		E_{14_16}	θ_G =W, act_G = *Through*
N_{15}	*Lmk* = *Door* $Dp_G \approx 260$ cm	E_{15_14}	θ_G =S, act_G = *Through*
N_{16}	*Lmk* = *Door* $Dp_G \approx 260$ cm	E_{16_1}	θ_G =W, act_G = *Goto*
		E_{16_14}	θ_G =E, act_G = *Through*

Fig. 6.6. Node and edge attributes of topological map for simulations

The node and edge attributes of the topological map are defined by the table form shown in Fig. 6.6. In the first column, N_1 to N_{16} symbolize nodes, like node ID. The second column displays the landmark feature *LmkF*, including landmark type *Lmk* and landmark depth Dp_G.

Dp_G is the approximate depth of a hollow goal, e.g. open doors and corridors, referred to as the safe distance, which is qualitatively given by the user, for going through a passable goal. Therefore, the Dp_G of a door is the relative distance from one side to other side of the door, e.g. the safe distance to the door is bigger than 130 cm, so $Dp_G \approx 2*130 = 260$ cm; and the Dp_G of a corridor is the relative distance smaller than the corridor length and bigger than half of this length, e.g. a corridor length approximates 920 cm, so $Dp_G \approx 620$ cm.

Notice that the Dp_G of walls or corners is always zero because mobile robots can never go through walls and corners.

The third column of the table contains the edge ID E_{i_j}, with $i \neq j$, and $i, j = 1 \ldots 16$. The last column is comprised of the action feature *ActF*, involving goal direction θ_G, $\theta_G \in \{N, E, S, W\}$, and motion action at the adjacent goal act_G, $act_G \in \{Goto, Though\}$.

6.1.3. Software structure

The software structure for the simulations is illustrated in Fig. 6.7. The simulation software comprises four main modules: the behavior-based controller module, the locomotion simulation module, the optical flow processor module, and the environment simulation module. The behavior-based controller module contains three sub-modules: the OFFIS, MOFIS and PAPFIS modules. All of these modules excluding the environment simulation module are written in Matlab source code (version 7.5).

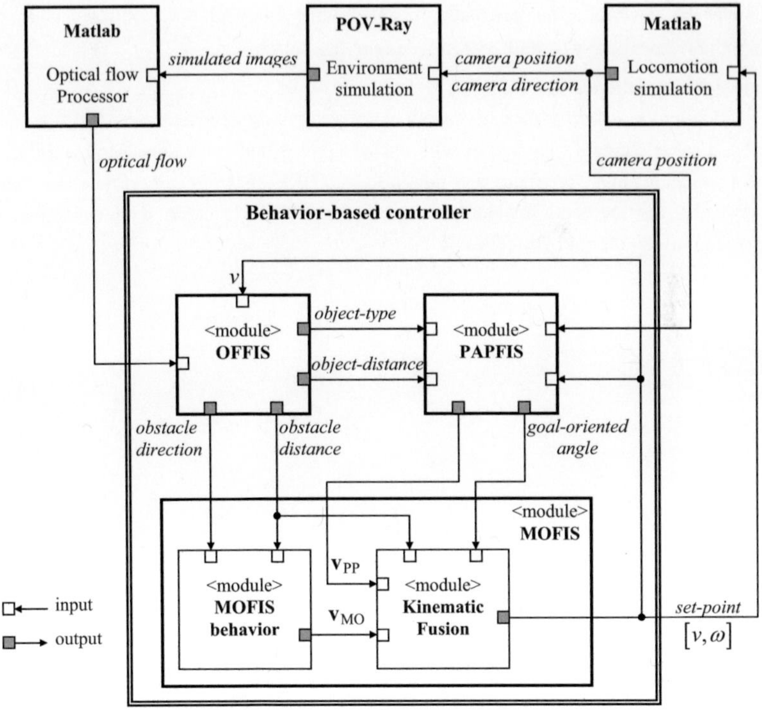

Fig. 6.7. Software structure of the control system for the simulations

As previously mentioned, the POV-Ray program reads the data in text files to generate simulated images. Therefore, the environment simulation module, written in POV-Ray, receives text files about camera position and direction from the locomotion simulation module; it then sends *simulated images* to the optical flow processor module.

The optical flow processor module extracts optical flow vectors from the simulated images and then delivers the translational components of *optical flow vectors* to the OFFIS module in the behavior-based controller.

The OFFIS module (see chapter 3) carries out the optical flow-based perception to determine 3D local information of the working environment about obstacles including *obstacle-direction* and *obstacle-distance*, about indoor landmarks including *object-type* and *object-distance*. Additionally, the OFFIS module uses the linear velocity v of the set-point, i.e. the output of the controller, to calculate fuzzy distances.

The PAPFIS module (see chapter 5) delivers a velocity command $\mathbf{v}_{\mathrm{PP}}$ for a goal-oriented behavior and a *goal-oriented angle* playing the role as decision-making information of the kinematic fusion.

The MOFIS module (see chapter 4) contains two sub-modules: MOFIS behavior and kinematic fusion. The MOFIS behavior module computes a velocity command $\mathbf{v}_{\mathrm{MO}}$ for a collision-free behavior. The kinematic fusion module executes a fuzzy-based cooperative coordination between $\mathbf{v}_{\mathrm{PP}}$ and $\mathbf{v}_{\mathrm{MO}}$ to deliver the set-point, $\mathbf{v} = [v, \omega]$, provided for the locomotion simulation module.

The locomotion simulation module calculates a new pose (position and direction) for the simulated robot (camera) that agrees with the set-point of the behavior-based controller. The camera position and direction are supported to the environment simulation module. Additionally, the locomotion module supports the PAPFIS module with camera position playing the role as the robot direction.

6.2. Simulation results

6.2.1. Simulation implementation with furniture

In the simulation implementations with furniture, named SIM-f, the simulated mobile robot performed the high-level navigation tasks of moving from one node to another node in the simulated environment, like an office, with some furniture, e.g. tables and chairs.

In the first simulation SIM-f-01, the end-to-end navigation task of the robot consisted of starting from node N_{15} in room R3 (see Fig. 6.4), travelling autonomously to node N_3 in room R1, and avoiding inherent furniture of the office.

The global path from node N_{15} to node N_3 was calculated automatically by the path planner of the PAPFIS module: *pth* = $\{N_{15}, N_{14}, N_{16}, N_1, N_2, N_3\}$.

The simulation results are shown in Fig. 6.8 by the ground truth of the driven trajectory of the simulated robot with several highlighted waypoint images ImS11, ImS12, ImS13 and ImS14.

Fig. 6.9 lists the optical flow interpretations of the OFFIS module with respect to the highlighted situations in Fig. 6.8.

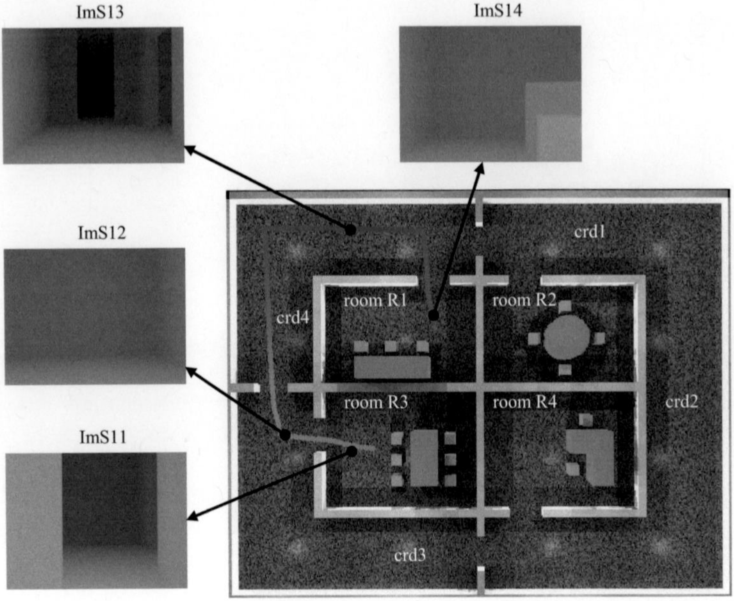

Fig. 6.8. Ground truth of the simulated robot in simulation SIM-f-01

Optical flow images (Hierarchical segmentation)	Optical flow interpretations (OFFIS outputs)
ImS11 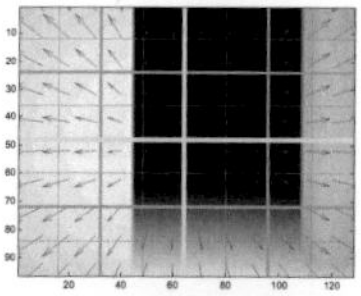	*obstacle-distance =* *0.607 0.843 0.892 0.698* *obstacle-direction = left* *TPO-distance =* *0.607 0.903 0.903 0.729* *TPO-type = door*
ImS12 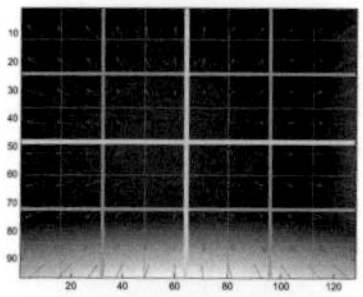	*obstacle_distance =* *0.727 0.892 0.892 0.726* *obstacle-direction = none* *TPO_distance =* *0.893 0.903 0.903 0.892* *TPO-type = corner*
ImS13 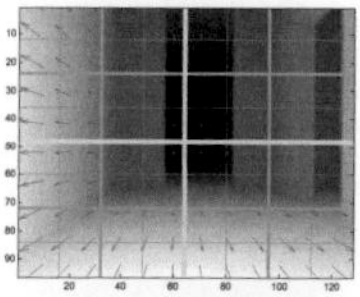	*obstacle_distance =* *0.715 0.892 0.892 0.727* *obstacle-direction = none* *TPO_distance =* *0.790 0.903 0.903 0.903* *TPO-type = door*
ImS14 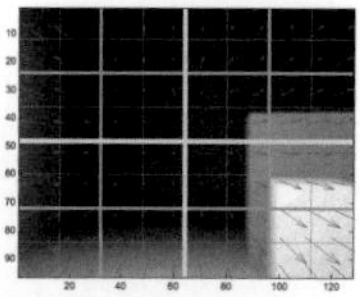	*obstacle_distance =* *0.718 0.892 0.903 0.536* *obstacle-direction = right* *TPO_distance =* *0.792 0.903 0.903 0.723* *TPO-type = corner*

Fig. 6.9. Optical flow interpretations of highlighted waypoint images in Fig. 6.8

In simulation SIM-f-02, the mobile robot executed another navigation task including starting from node N_3 in side room R1 and moving automatically to node N_{10} inside the room R4.

The path planner of the PAPFIS module calculated also automatically a global path from N_3 to N_{10}. Notice that, from N_3 to N_{10}, there were two available paths of the same distance: the path over corridors crd1 and crd2 contains 8 nodes $\{N_3, N_2, N_4, N_6, N_7, N_8, N_9, N_{10}\}$, and the other path over corridors crd1 and crd4 has 10 nodes $\{N_3, N_2, N_1, N_{16}, N_{14}, N_{13}, N_{12}, N_{11}, N_9, N_{10}\}$. So the path planner selected the first path with a smaller number of nodes.

The result of simulation SIM-f-02 is illustrated in Fig. 6.10 by the ground truth of the driven trajectory of the simulated robot with several highlighted waypoint images ImS21, ImS22, ImS23, ImS24, ImS25 and ImS26.

Fig. 6.11 displays the optical flow interpretations of the OFFIS module with respect to the highlighted situations in Fig. 6.10.

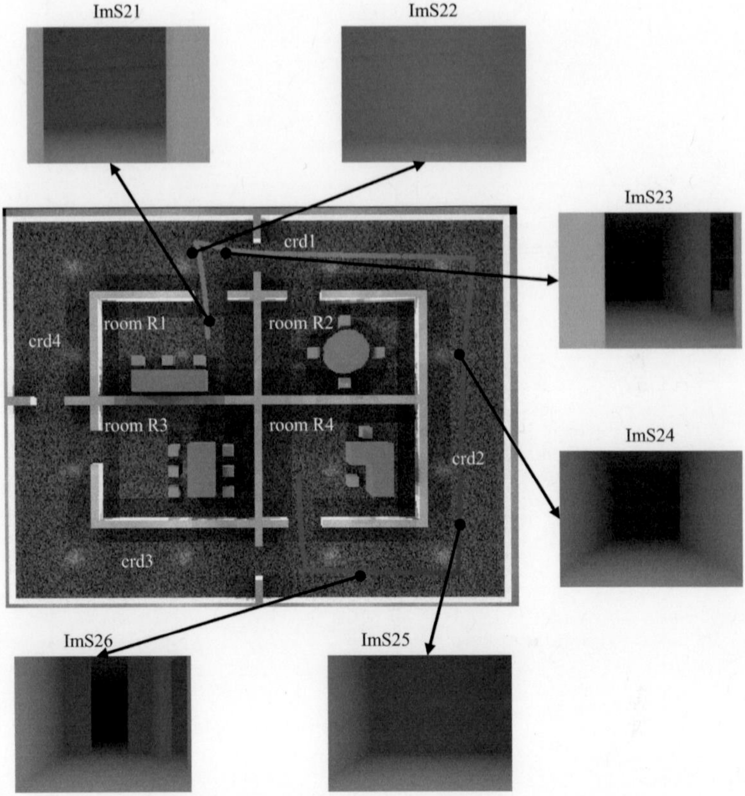

Fig. 6.10. Ground truth of the simulated robot in simulation SIM-f-02

Optical flow images (Hierarchical segmentation)	Optical flow interpretations (OFFIS outputs)
ImS21 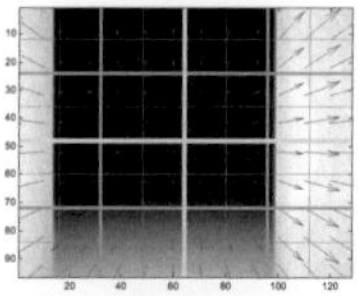	*obstacle-distance =* *0.623 0.892 0.892 0.543* *obstacle-direction = right* *TPO-distance =* *0.714 0.903 0.903 0.546* *TPO-type = door*
ImS22 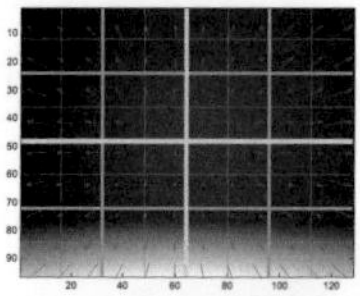	*obstacle-distance =* *0.727 0.892 0.892 0.725* *obstacle-direction = none* *TPO-distance =* *0.895 0.903 0.903 0.889* *TPO-type = wall*
ImS23 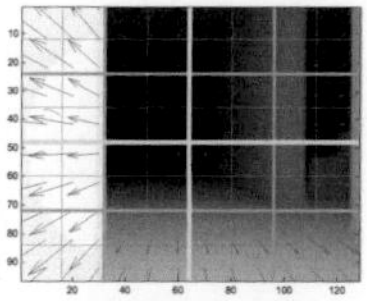	*obstacle-distance =* *0.394 0.892 0.892 0.720* *obstacle-direction = left* *TPO-distance =* *0.395 0.903 0.903 0.903* *TPO-type = door*
ImS24 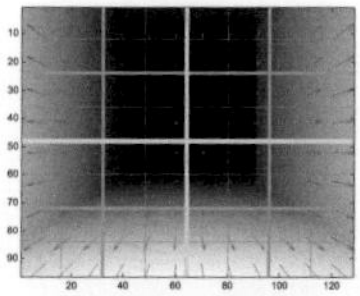	*obstacle-distance =* *0.724 0.892 0.892 0.723* *obstacle-direction = none* *TPO-distance =* *0.896 0.903 0.903 0.890* *TPO-type = corridor*
ImS25 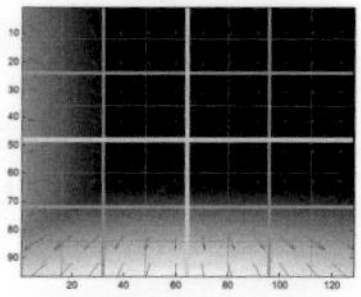	*obstacle-distance =* *0.725 0.892 0.892 0.729* *obstacle-direction = none* *TPO-distance =* *0.897 0.903 0.903 0.903* *TPO-type = corner*
ImS26 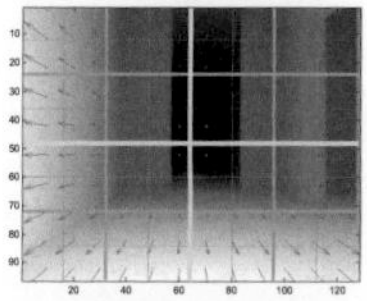	*obstacle-distance =* *0.714 0.892 0.892 0.728* *obstacle-direction = none* *TPO-distance =* *0.780 0.903 0.903 0.903* *TPO-type = door*

Fig. 6.11. Optical flow interpretations of highlighted waypoint images in Fig. 6.10

6.2.2. Simulation implementation with static obstacles

In these simulation implementations, symbolized by SIM-s, the navigation tasks for the robot were more difficult than those in the simulation SIM-f by adding the task of avoiding all unknown static obstacles randomly arranged along the corridors shown in Fig. 6.12.

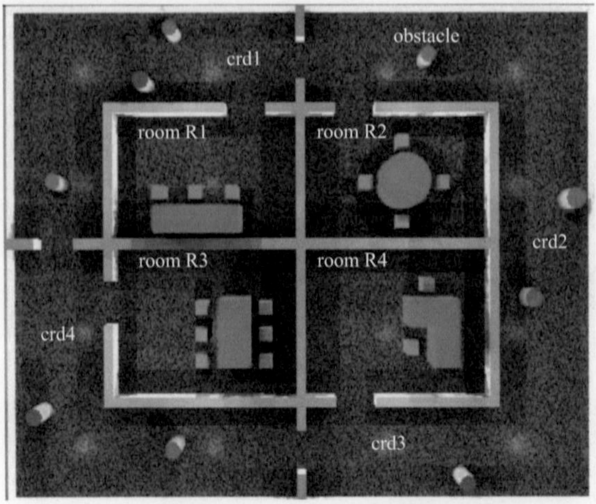

Fig. 6.12. Obstacle arrangement for simulations SIM-s

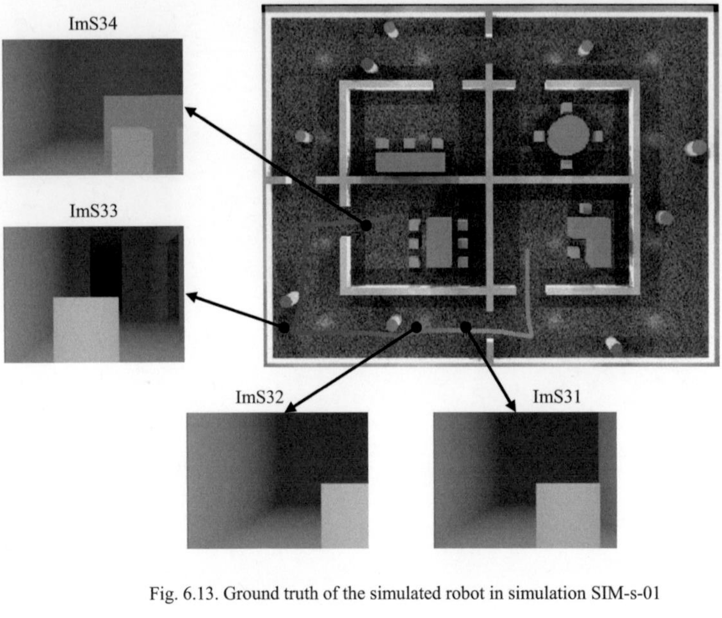

Fig. 6.13. Ground truth of the simulated robot in simulation SIM-s-01

In the simulation SIM-s-01, the high-level navigation tasks of the mobile robot were to move from node N_{10} (inside room R4) to node N_{15} (inside room R3), and avoid the static obstacles on the road.

The shortest global path computed automatically by the global planner on the topological map was $\{ N_{10}, N_9, N_{11}, N_{12}, N_{13}, N_{14}, N_{15}\}$.

The result of simulation SIM-s-01 is illustrated in Fig. 6.13 by the ground truth of the driven trajectory of the simulated robot with several highlighted waypoint images ImS31, ImS32, ImS33 and ImS34.

Fig. 6.14 shows the optical flow interpretations of the OFFIS module with respect to the highlighted situations in Fig. 6.13.

Optical flow images (Hierarchical segmentation)	**Optical flow interpretations (OFFIS outputs)**
ImS31 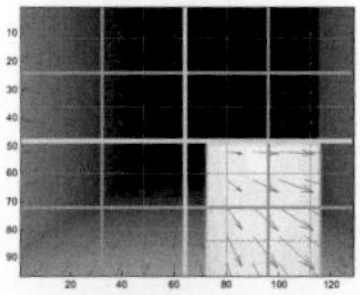	*obstacle-distance =* *0.718 0.892 0.609 0.563* *obstacle-direction = right* *TPO-distance =* *0.831 0.903 0.903 0.903* *TPO-type = corridor*
ImS32 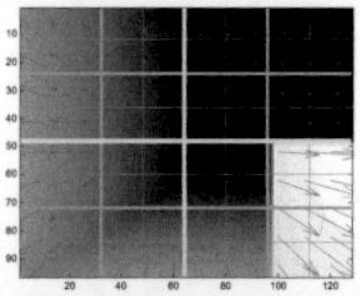	*obstacle-distance =* *0.663 0.892 0.903 0.331* *obstacle-direction = right* *TPO-distance =* *0.684 0.903 0.903 0.903* *TPO-type = corner*
ImS33 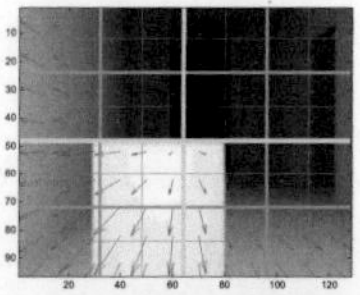	*obstacle-distance =* *0.739 0.478 0.643 0.903* *obstacle-direction = left* *TPO-distance =* *0.739 0.903 0.903 0.903* *TPO-type = door*
ImS34 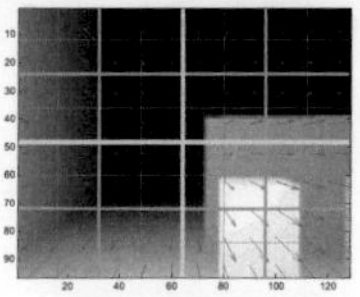	*obstacle-distance =* *0.719 0.892 0.719 0.619* *obstacle-direction = right* *TPO-distance =* *0.838 0.903 0.903 0.893* *TPO-type = corner*

Fig. 6.14. Optical flow interpretations of highlighted waypoint images in Fig. 6.13

Different to simulation SIM-s-01, in simulation SIM-s-02, the mobile robot had to travel from node N_{15} (inside room R3) to the end point was node N_5 (inside the room R2). The robot also avoided the obstacles on the road.

The final result of simulation SIM-s-02 is recorded by the ground truth of the driven trajectory of the simulated mobile robot, and illustrated by some highlight waypoint images ImS41, ImS42, ImS43 and ImS44 in Fig. 6.15. The optical flow interpretations of the OFFIS module with respect to the highlighted images in Fig. 6.15 are listed in Fig. 6.16.

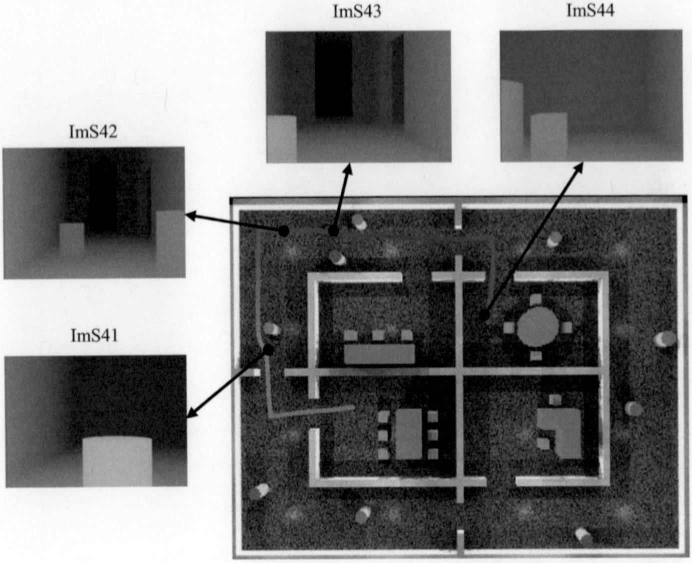

Fig. 6.15. Ground truth of the simulated robot in simulation SIM-s-02

Optical flow images (Hierarchical segmentation)	Optical flow interpretations (OFFIS outputs)
ImS41 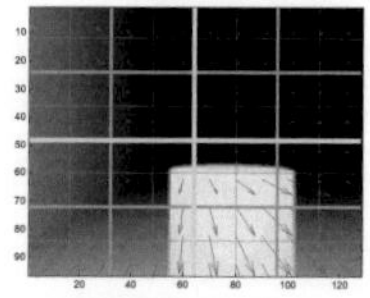	*obstacle-distance =* *0.837 0.707 0.467 0.903* *obstacle-direction = right* *TPO-distance =* *0.837 0.903 0.903 0.903* *TPO-type = corner*
ImS42 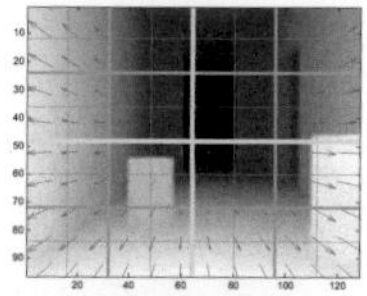	*obstacle-distance =* *0.710 0.891 0.892 0.681* *obstacle-direction = right* *TPO-distance =* *0.745 0.903 0.903 0.902* *TPO-type = corridor*
ImS43 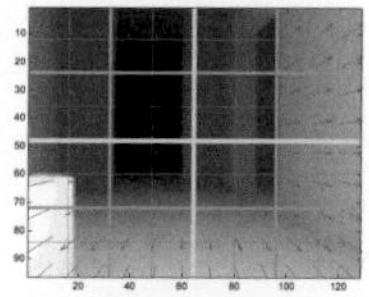	*obstacle-distance =* *0.535 0.892 0.892 0.713* *obstacle-direction = left* *TPO-distance =* *0.903 0.903 0.903 0.750* *TPO-type = door*
ImS44 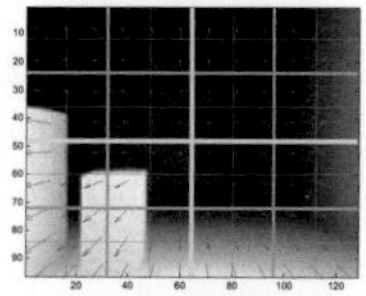	*obstacle-distance =* *0.641 0.841 0.892 0.722* *obstacle-direction = left* *TPO-distance =* *0.866 0.903 0.903 0.888* *TPO-type = corner*

Fig. 6.16. Typical Optical flow interpretations of highlighted waypoint images in Fig. 6.15

6.2.3. Simulation implementation with narrow areas

In reality, mobile robot navigation tasks can be done in narrow areas, for example, inside small rooms or through small regions with a large number of obstacles. For this reason, in the following simulations SIM-n, the robot worked in narrow areas.

For the simulation implementations, called SIM-n, the simulated environment, illustrated in Fig, 6.17, contained more static obstacles. Particularly, the static obstacles were arranged randomly with a high density along the corridors crd3 and crd4.

The navigation tasks in the simulations SIM-n were more complex than those in other previous simulations because of the new tasks: starting in the corner of a room, passing through risky corridors with obstacle density as illustrated in Fig. 6.17, and finishing in the corner of another room. Furthermore, after reaching the target, the robot returned to the start corner in the start room.

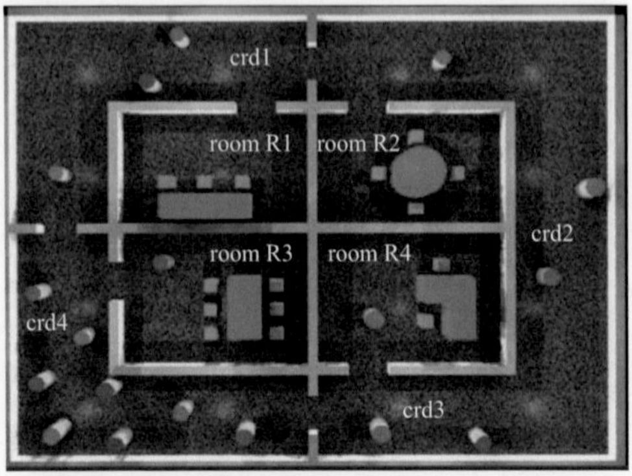

Fig. 6.17. High density of obstacles in simulations SIM-n

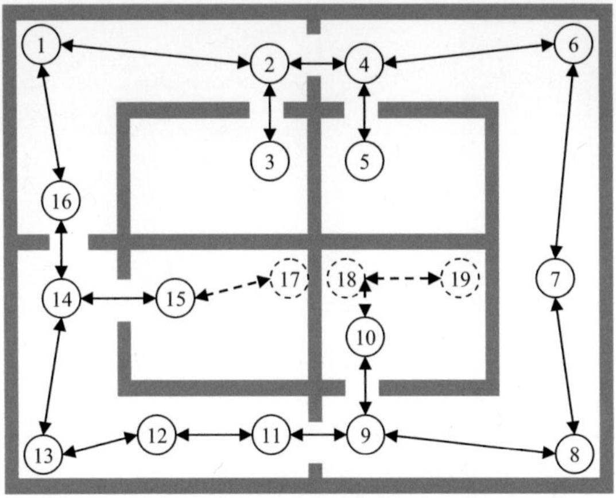

Fig. 6.18. Expansion of topological map for simulations SIM-n

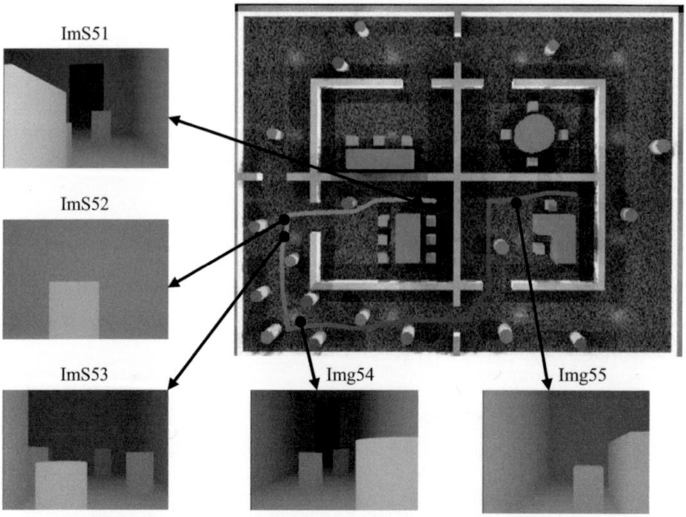

Fig. 6.19. Ground truth of the simulated robot in simulation SIM-n-01

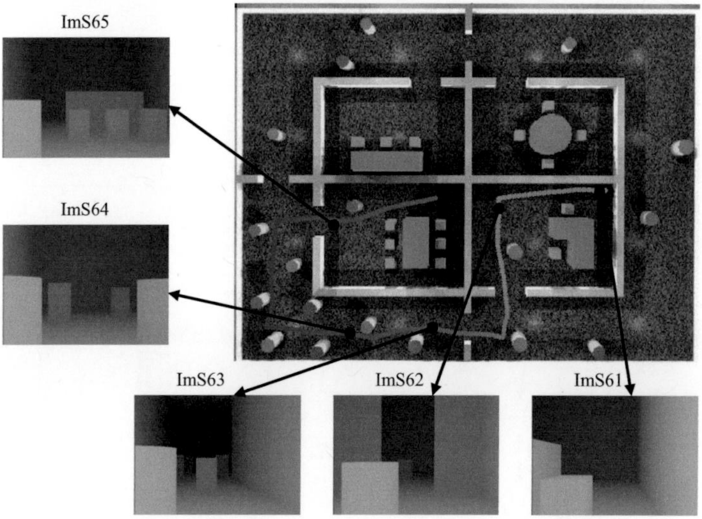

Fig. 6.20. Ground truth of the simulated robot in simulation SIM-n-02

Due to the tasks involving some corners in the rooms, an expansion of the topological map by adding some nodes at the corners is required. Such an expanded topological map is illustrated in Fig. 6.18 by adding nodes N_{17}, N_{18} and N_{19} to the corners in room R3 and R4.

The navigation task of the robot in the simulations SIM-n included moving from node N_{17} (inside room R3) to node N_{19} (inside room R4) and then returning to node N_{17}.

The results of simulations SIM-n were divided into two periods: the away movement from node N_{17} to node N_{19}, and the home movement from node N_{19} to node N_{17}. That means simulations SIM-n were referred to as two separate simulations, SIM-n-01 and SIM-n-02.

Optical flow images (Hierarchical segmentation)	**Optical flow interpretations (OFFIS outputs)**
ImS51 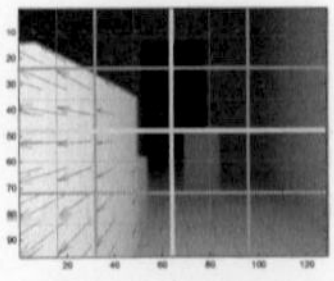	*obstacle-distance =* *0.298 0.703 0.892 0.721* *obstacle-direction = left* *TPO-distance =* *0.299 0.903 0.903 0.892* *TPO-type = corridor*
ImS52 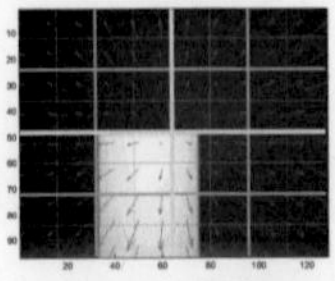	*obstacle-distance =* *0.718 0.562 0.716 0.680* *obstacle-direction = left* *TPO-distance =* *0.718 0.889 0.885 0.698* *TPO-type = wall*
ImS53 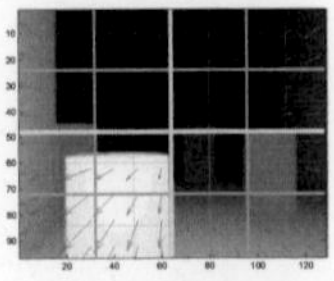	*obstacle-distance =* *0.835 0.568 0.892 0.722* *obstacle-direction = left* *TPO-distance =* *0.835 0.903 0.903 0.903* *TPO-type = corridor*
ImS54 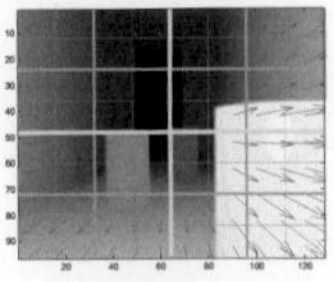	*obstacle-distance =* *0.728 0.891 0.659 0.350* *obstacle-direction = right* *TPO-distance =* *0.903 0.903 0.903 0.514* *TPO-type = corridor*
ImS55 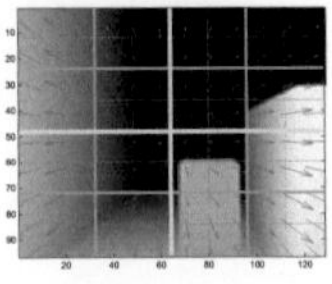	*obstacle-distance =* *0.622 0.892 0.738 0.531* *obstacle-direction = right* *TPO-distance =* *0.629 0.903 0.903 0.600* *TPO-type = corner*

Fig. 6.21. Optical flow interpretations of highlighted waypoint images in Fig. 6.19

Optical flow images (Hierarchical segmentation)	Optical flow interpretations (OFFIS outputs)
ImS61 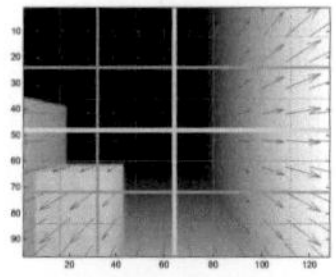	*obstacle-distance =* *0.544 0.773 0.889 0.446* *obstacle-direction = left* *TPO-distance =* *0.900 0.903 0.903 0.446* *TPO-type = corner*
ImS62 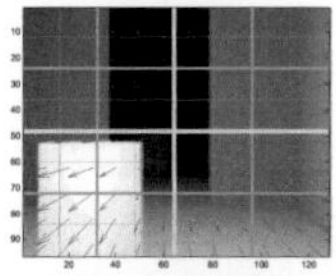	*obstacle-distance =* *0.557 0.719 0.892 0.720* *obstacle-direction = left* *TPO-distance =* *0.895 0.903 0.903 0.813* *TPO-type = door*
ImS63 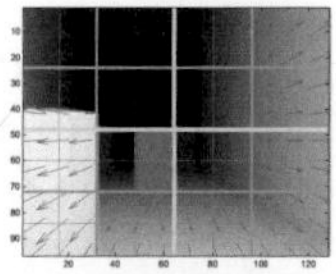	*obstacle-distance =* *0.502 0.892 0.892 0.707* *obstacle-direction = left* *TPO-distance =* *0.742 0.903 0.903 0.727* *TPO-type = corridor*
ImS64 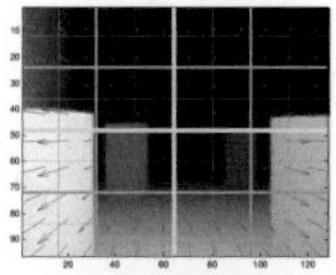	*obstacle-distance =* *0.519 0.892 0.903 0.622* *obstacle-direction = left* *TPO-distance =* *0.718 0.903 0.903 0.903* *TPO-type = corner*
ImS65 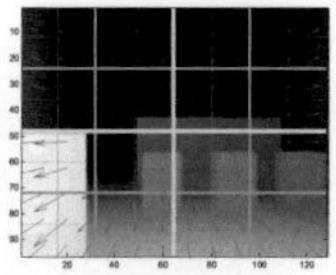	*obstacle-distance =* *0.486 0.891 0.891 0.724* *obstacle-direction = left* *TPO-distance =* *0.903 0.903 0.903 0.903* *TPO-type = corner*

Fig. 6.22. Optical flow interpretations of highlighted waypoint images in Fig. 6.20

The ground truths of the driven trajectory of the simulated robot in simulations SIM-n-01 and SIM-n-02 are illustrated in Fig. 6.19 and Fig 6.20 with two set of highlighted waypoint images {ImS51, ImS52, ImS53, ImS54, ImS55} and {ImS61, ImS62, ImS63, ImS64, ImS65}, respectively.

Fig. 6.21 and Fig. 6.22 illustrate the optical flow interpretations of the OFFIS module with respect to the highlighted situations in Fig. 6.19 and Fig. 6.20, respectively.

6.2.4. Simulation implementation with dynamic obstacles

In the subsequent simulation implementations, denoted SIM-d, the mobile robot encountered dynamic obstacles on the road. The movements of the dynamic obstacles are (in yellow) illustrated in Fig. 6.23. Some dynamic obstacles moved linearly, while others moved casually.

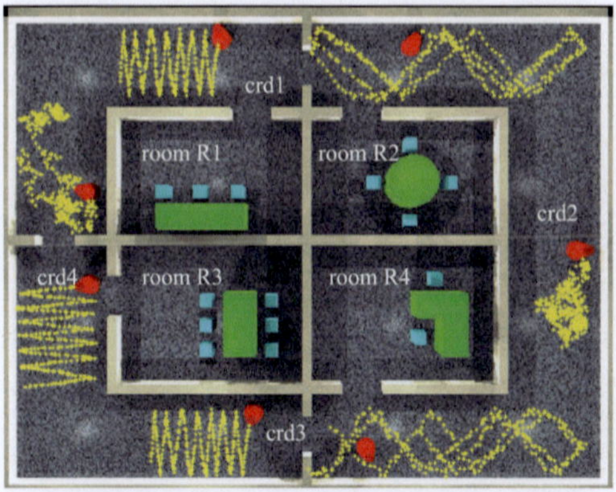

Fig. 6.23. Paths of dynamic obstacles in simulations SIM-d

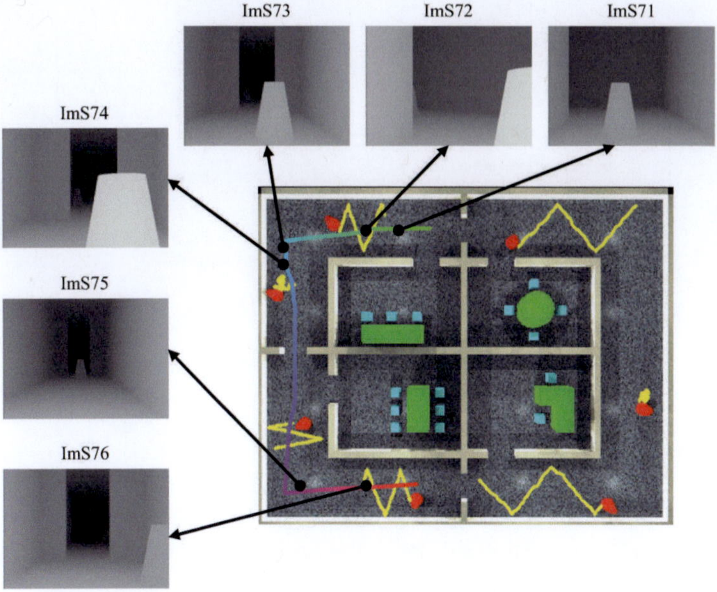

Fig. 6.24. Ground truth of the simulated robot in simulation SIM-d-01

Optical flow images (Hierarchical segmentation)	Optical flow interpretations (OFFIS outputs)
ImS71 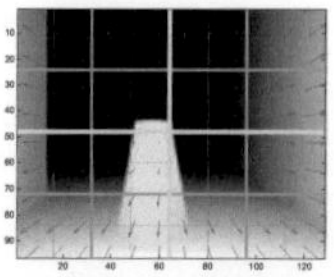	*obstacle-distance =* *0.726 0.855 0.890 0.720* *obstacle-direction = none* *TPO-distance =* *0.903 0.903 0.903 0.887* *TPO-type = corridor*
ImS72 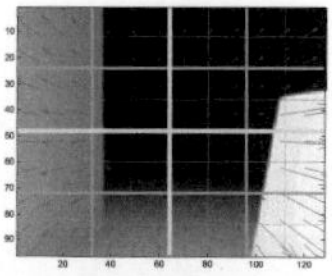	*obstacle-distance =* *0.615 0.891 0.892 0.298* *obstacle-direction = right* *TPO-distance =* *0.617 0.903 0.903 0.531* *TPO-type = corner*
ImS73 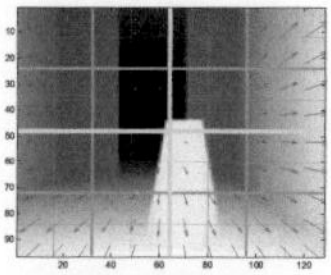	*obstacle-distance =* *0.729 0.888 0.850 0.710* *obstacle-direction = none* *TPO-distance =* *0.903 0.903 0.903 0.747* *TPO-type = corridor*
ImS74 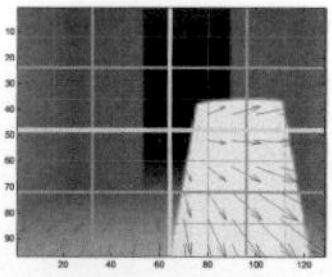	*obstacle-distance =* *0.726 0.892 0.441 0.400* *obstacle-direction = right* *TPO-distance =* *0.897 0.903 0.903 0.705* *TPO-type = door*
ImS75 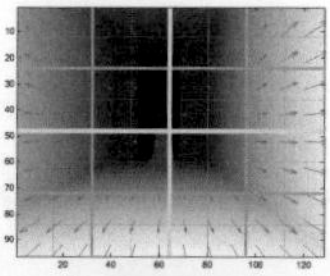	*obstacle-distance =* *0.728 0.892 0.892 0.707* *obstacle-direction = none* *TPO-distance =* *0.903 0.903 0.903 0.733* *TPO-type = corridor*
ImS76 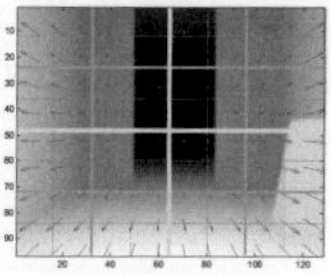	*obstacle-distance =* *0.721 0.892 0.892 0.703* *obstacle-direction = none* *TPO-distance =* *0.887 0.903 0.903 0.892* *TPO-type = door*

Fig. 6.25. Optical flow interpretations of highlighted waypoint images in Fig. 6.24

In simulation SIM-d-01, the navigation tasks of the mobile robot were to start from node N_2 (the door of corridor crd1), finish at node N_{11} (the door of corridor crd3), and avoid the dynamic obstacles in corridors crd1, crd3, and crd4. Noticeably, in this simulation, the maximum velocity of the dynamic obstacles was approximately half of the maximum velocity of the mobile robot.

The final result of simulation SIM-d-01 is recorded by the ground truth of the driven trajectory of the simulated mobile robot, illustrated in Fig. 6.24, and displayed by some highlight waypoint images ImS71, ImS72, ImS73, ImS74, ImS75 and ImS76. The optical flow interpretations with respect to these highlighted images are shown in Fig. 6.25.

In simulation SIM-d-02, the mobile robot executed the navigation task including starting from node N_9 (the door of corridor crd3), finishing at node N_2 (the door of corridor crd1), and avoiding the dynamic obstacles on the road.

The important difference of simulation SIM-d-02 to simulation SIM-d-01 is that in simulation SIM-d-02 the maximum velocity of the dynamic obstacles was approximately that of the maximum velocity of the mobile robot. This means the dynamic obstacles in this simulation are faster.

The simulation results of simulation SIM-d-02 are displayed in Fig. 6.26 by the ground truth of the driven trajectory of the simulated mobile robot with several highlighted waypoint images ImS81, ImS82, ImS83, ImS84, ImS85 and ImS86.

Fig. 6.27 illustrates the optical flow interpretations of the OFFIS module with respect to the highlighted situations in Fig. 6.26.

The collision-freeness of robot motion can be deduced from the continuous ground truth paths from the start point to the end point.

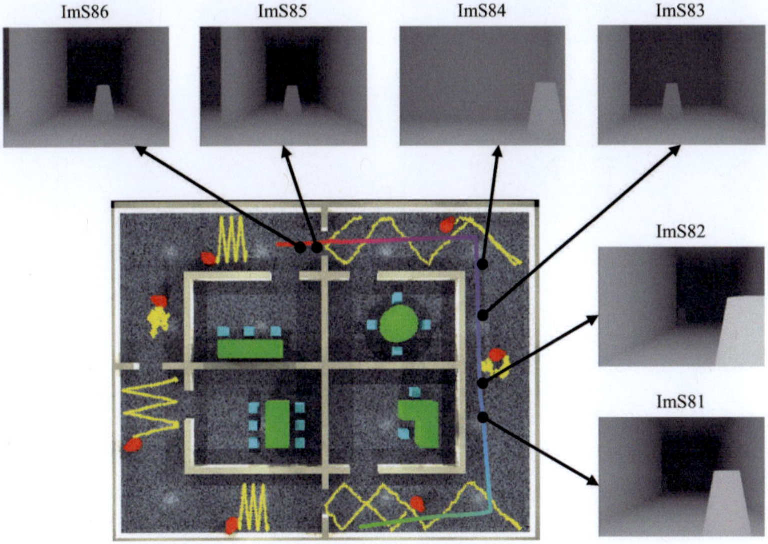

Fig. 6.26. Ground truth of the simulated robot in simulation SIM-d-02

Optical flow images (Hierarchical segmentation)	**Optical flow interpretations (OFFIS outputs)**
ImS81 	*obstacle-distance =* *0.724 0.892 0.771 0.644* *obstacle-direction = right* *TPO-distance =* *0.893 0.903 0.903 0.902* *TPO-type = corridor*
ImS82 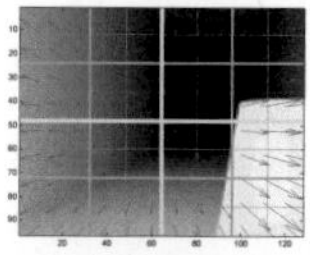	*obstacle-distance =* *0.712 0.892 0.770 0.386* *obstacle-direction = right* *TPO-distance =* *0.747 0.903 0.903 0.713* *TPO-type = corridor*
ImS83 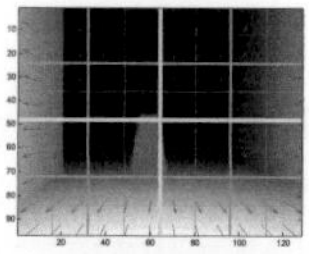	*obstacle-distance =* *0.724 0.892 0.892 0.728* *obstacle-direction = none* *TPO-distance =* *0.900 0.903 0.903 0.903* *TPO-type = corridor*
ImS84 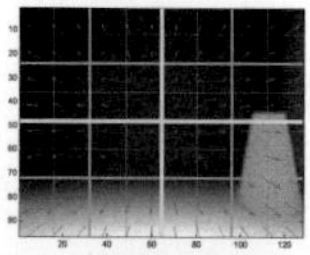	*obstacle-distance =* *0.729 0.892 0.892 0.707* *obstacle-direction = none* *TPO-distance =* *0.903 0.903 0.903 0.898* *TPO-type = wall*
ImS85 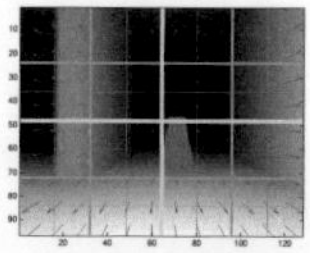	*obstacle-distance =* *0.728 0.892 0.892 0.727* *obstacle-direction = none* *TPO-distance =* *0.903 0.903 0.903 0.903* *TPO-type = door*
ImS86 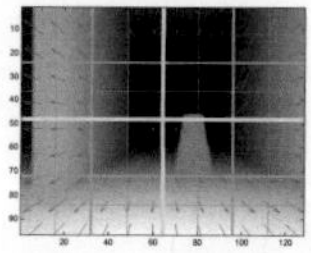	*obstacle-distance =* *0.722 0.892 0.892 0.728* *obstacle-direction = none* *TPO-distance =* *0.889 0.903 0.903 0.903* *TPO-type = corridor*

Fig. 6.27. Optical flow interpretations of highlighted waypoint images in Fig. 6.26

6.3. Discussion

The simulation results exhibit the stability of the optical flow perception in the various scenarios. The proposed approach for optical flow perception allows a mobile robot to execute a simple landmark-based localization without having to match observed features and expected features. This means the mobile robot does not need a big memory to store the complex database of landmark features.

Additionally, the results of the simulations demonstrate the smooth operation of the mobile robot navigation in the working environment, containing static and dynamic obstacles. Note that, in the simulations SIM-d, the mobile robot deals with dynamic obstacles via the static obstacle avoidance strategy because it has no information about the kinematics of the dynamic obstacles. For this reason, a collision may occur if dynamic obstacles move too fast. In other words, the faster the dynamic obstacles move, is the higher the collision probability will be.

Moreover, the simulation results demonstrate the stable operation of the behavior-based control system with topological path planning based on the topological map and gyroscope. This indicates that the topological path planning is able to be used for mobile robot guidance in circumstances with less metric information. Additionally, the results of simulations SIM-n present the flexibility of topological path planning in the situations requiring an adjustment of the topological map for a new navigation task.

Noticeably, through all simulations, the initialization of the robot position does not demand exact coordinates, but it should be next to a start node. This makes the robot operation simpler. Additionally, the coordinates of robot position are not required. This means the problem of accumulated error position is no longer an issue.

Critically, simulation represents an ideal environment where the robustness of the system operation against noise is not tested. That means the environmental conditions of simulations are reputedly perfect. What will happen when the mobile robot works in real environments? The answer for this question will be solved by the experiments in the next chapter.

Chapter 7
Experimental test and verification

This chapter presents the results of the experiments under most realistic conditions. These experiments have been performed in a real office environment with real camera images to prove the possible performances of the optical flow perception and the robust operation of the behavior-based control system. The experimental results are illustrated by the ground truths of a real mobile robot with several highlighted waypoint images.

7.1. Experiment setups

7.1.1. Experiment environment

The real environment for the experiments was an office area of the institute of automation (IfA) at the Technische Universität Dresden (TU Dresden). The experimental environment comprised of several rooms with figures S4 to S9, doors with the symbols dr4 to dr9, corridor crd49, and a staircase shown in Fig. 7.1. There is also some inherent furniture, such as chairs, tables, wardrobe and dustbins.

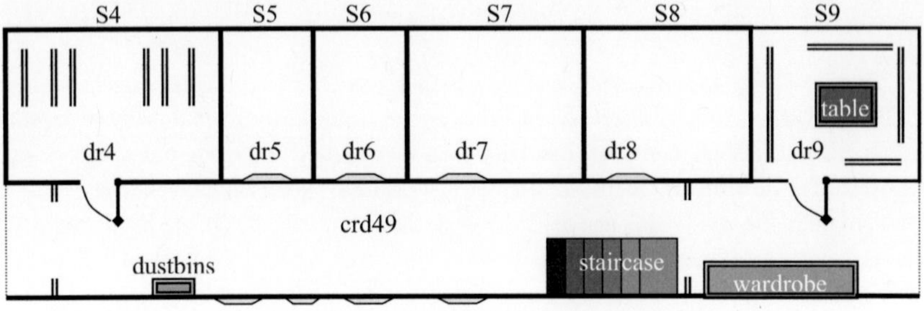

Fig. 7.1. Building plan of the experimental environment

7.1.2. Hardware

The robot platform used for all experiments is a mobile robot IfAbot built by the robotics lab of the Institute of Automation (IfA) at the TU Dresden.

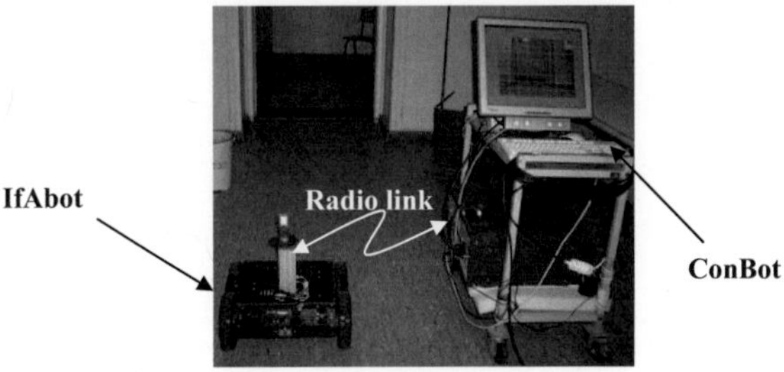

Fig. 7.2. Hardware of the experiments

Fig. 7.2 displays the hardware for the experiments: the IfAbot platform (on the left hand) and a ConBot platform (on the right hand) carrying some peripheral devices that were necessary for control motion of the IfAbot. These platforms communicate via a radio link.

a) Features of the IfAbot mobile robot

IfAbot is a skid-steered mobile robot, illustrated in Fig. 7.3, containing various sensors, such as sonar sensors, laser sensors, gyroscope, odometry and digital camera.

Fig. 7.3. IfAbot platform

Throughout the experiments, the IfAbot mobile robot used the digital camera Logitech C310 HD with 1280 x 720 screen resolution as the main sensor. Additionally, it used the gyroscope ADXRS300 to measure robot direction as necessary for topological path planning.

According to (Modes, 2010), the IfAbot platform has three sub-modules, including the PC-controller, the robot-plate, and the IMU-plate shown in Fig. 7.4. The IfAbot platform also has some standard interfaces, such as RS232, USB and WLAN, for internal and external connections.

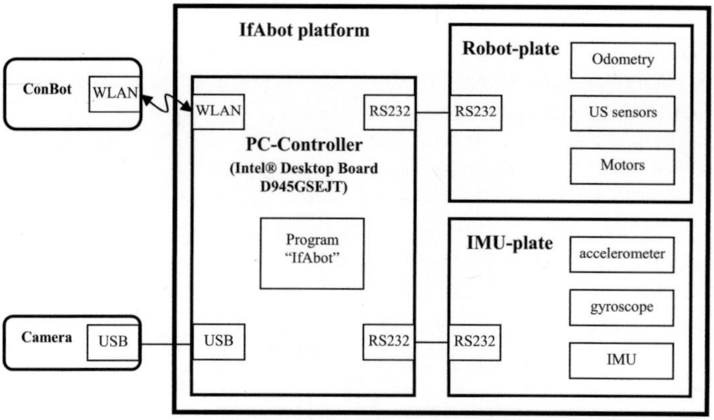

Fig. 7.4. Architecture diagram of IfAbot (Modes, 2010)

More details about the IfAbot specification, including the driving system, control architecture and software plan, are described in (Modes, 2010) and (Schnellhorn, 2009).

b) Features of the ConBot platform

The ConBot platform carries a GPU to speed up image processing, a CPU to operate the behavior-based controller program, and a monitor to feedback and visualize the experiment results.

The GPU is a powerful graphics engine because of the application of graphics pipelines that execute parallel computation. This increases the speed of image.

More details about the specifications of the GPU are described in (Kimmer, 2009).

7.1.3. Software structure

The modular software structure for the experiments is shown in Fig. 7.5, where the MOFIS module is depicted with two sub-modules, including the MOFIS behavior and the kinematic fusion. The kinematic fusion module makes a fusion of the velocities $\mathbf{v}_{PP}$ and $\mathbf{v}_{MO}$ of the PAPFIS and MOFIS behaviors, respectively.

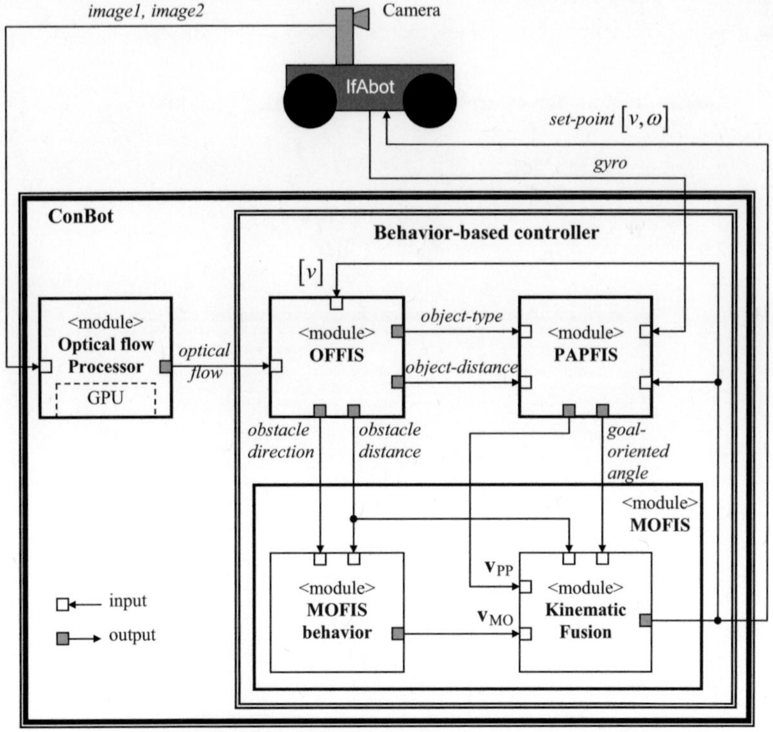

Fig. 7.5. Software structure of the navigation system for the experiments

The IfAbot grabs two images from the camera at two different positions and send the grabbed images to the ConBot platform via the radio link. The optical flow processor in the ConBot platform produces an optical flow field for each grabbed image pair by the GPU NVIDIA GeForce GTX 260 1.24GHz through a 2D correlation block matching algorithm with segment size 32 x 32 pixels. Once completed, the optical flow processor delivers the translational components of optical flow to the OFFIS.

The OFFIS module has four outputs, including *obstacle-distance*, *obstacle-direction*, *object-distance*, and *object-type*, which interface with the motion controller MOFIS and the path planner PAPFIS as shown in Fig. 7.5. The software for the experiments is written in Matlab source code such as m-files, and it uses the fuzzy logic toolbox, e.g. for the fuzzy distance calculation and the fuzzy-based interpretation in the OFFIS module. These algorithms are executed on the computer Intel(R) Pentium(R) 4 CPU 3.40GHz.

The PAPFIS module delivers a velocity command $\mathbf{v}_{PP}$ of a goal-oriented behavior to the MOFIS behavior module, and a goal-oriented angle to the kinematic fusion module in the MOFIS module. The MOFIS behavior module calculates a velocity command $\mathbf{v}_{MO}$ for a collision-free behavior based on the obstacle information. The kinematic fusion module executes the fuzzy-based kinematic fusion between the two competitive velocities $\mathbf{v}_{MO}$ and $\mathbf{v}_{PP}$ to provide the set-point, $\mathbf{v} = [v,\ \omega]$, for the locomotion of the IfAbot.

7.2. Experiment of motion control without path planning EXP-MO

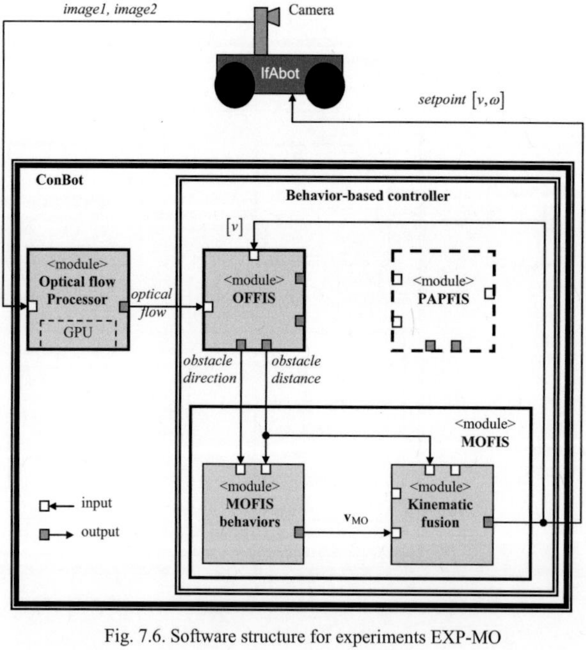

Fig. 7.6. Software structure for experiments EXP-MO

Fig. 7.7. Scene of experiment EXP-MO1

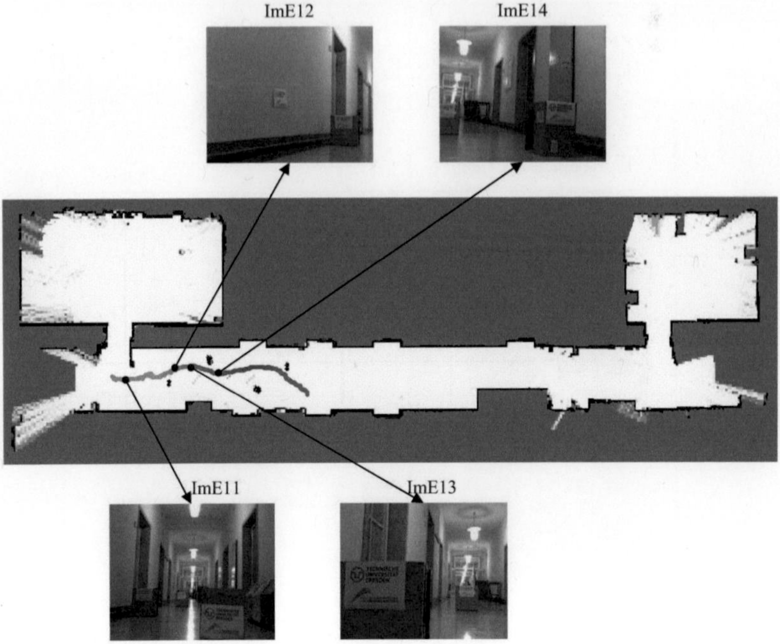

Fig. 7.8. Ground truth of the driven trajectory of the IfAbot in experiment EXP-MO1

The objective of these experiments, symbolized by EXP-MO, was to test the operation of the behavior-based controller *with* MOFIS behaviors and *without* PAPFIS behaviors. That means the topological path planning was not available in these experiments.

The high-level navigation tasks of the IfAbot in the experiments EXP-MO concerned travelling through a 3-by-10-meter area, displayed in Fig. 7.7, and avoiding the unknown artificial obstacles randomly arranged in this area.

Fig. 7.6 illustrates the software structure for experiments EXP-MO, which had been modified from the software structure shown in Fig. 7.5 by removing the PAPFIS module.

In the first experiment EXP-MO1, the obstacles comprised of the inherent obstacles, e.g. the walls and dustbins, and the artificial obstacles, e.g. the cartons randomly settled along the corridor, illustrated in Fig. 7.7.

The robot successfully made it through the risky area without collision. The result of the experiment EXP-MO1 is illustrated in Fig. 7.8 by the ground truth of the driven way of the IfAbot with several highlighted waypoint images ImE11, ImE12, ImE13 and ImE14.

Fig. 7.9 exhibits the OFFIS optical flow interpretations with respect to the highlighted situations in Fig. 7.8.

What will happen if a new obstacle appears in the path of the robot?

Optical flow images (Hierarchical segmentation)	**Optical flow interpretations (OFFIS outputs)**

ImE11
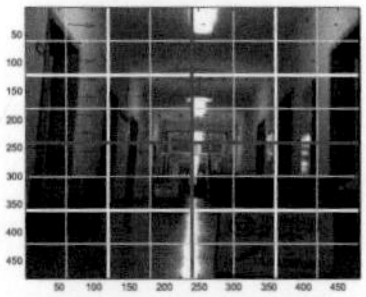
	`obstacle-distance =`
	`0.903 0.903 0.903 0.701`
	`obstacle-direction = right`

ImE12
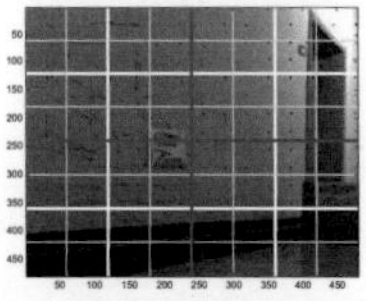
	`obstacle-distance =`
	`0.718 0.903 0.903 0.903`
	`obstacle-direction = left`

ImE13
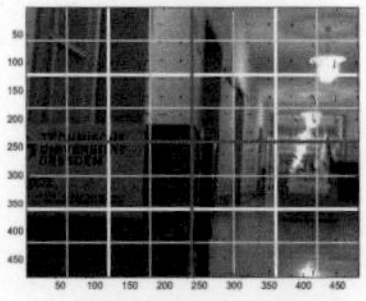
	`obstacle-distance =`
	`0.127 0.471 0.903 0.903`
	`obstacle-direction = left`

ImE14
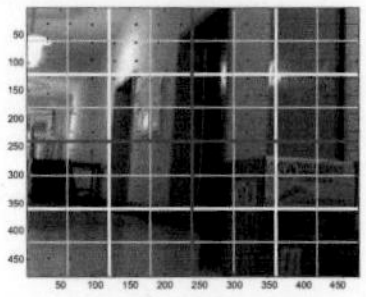
	`obstacle-distance =`
	`0.903 0.903 0.903 0.566`
	`obstacle-direction = right`

Fig. 7.9. Optical flow interpretations of highlighted waypoint images in Fig. 7.8

Fig. 7.10. Scene of experiment EXP MO2

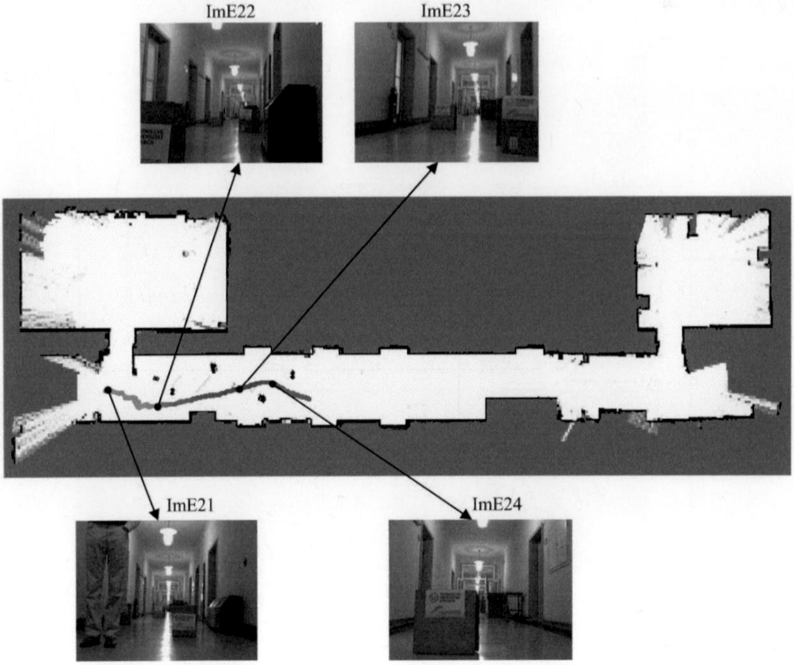

Fig. 7.11. Ground truth of the driven trajectory of the IfAbot in experiment EXP-MO2

To answer this question, experiment EXP-MO2 was implemented where a person stood in the path of the robot, specifically, in front of it and to the left, as displayed in Fig. 7.10. This situation is to check the reliable capability of obstacle avoidance of the IfAbot.

The robot automatically computed a different path in order to avoid the new obstacle. The result of the experiment EXP-MO2 is demonstrated by the ground truth of the driven trajectory of the IfAbot in Fig. 7.11 with some highlighted waypoint images ImE21, ImE22, ImE23 and ImE24.

Fig. 7.12 illustrates the optical flow interpretations of the OFFIS module with respect to the highlighted situations in Fig. 7.11.

Optical flow images (Hierarchical segmentation)	**Optical flow interpretations (OFFIS outputs)**
ImE21 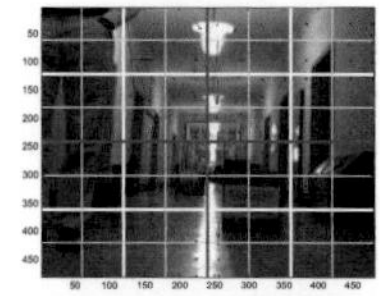	`obstacle-distance = 0.397 0.903 0.903 0.903 obstacle-direction = left`
ImE22 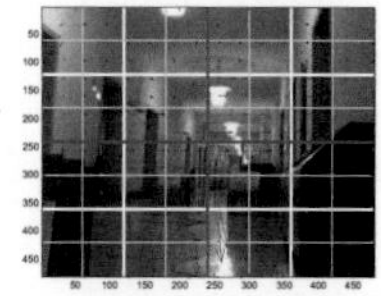	`obstacle-distance = 0.186 0.903 0.454 0.897 obstacle-direction = left`
ImE23 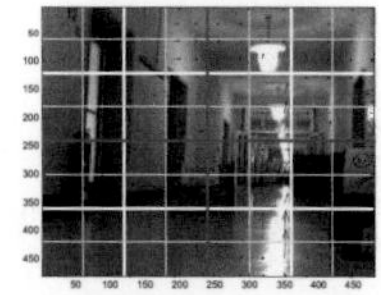	`obstacle-distance = 0.903 0.889 0.790 0.732 obstacle-direction = right`
ImE24 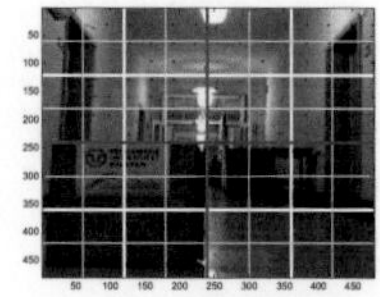	`obstacle-distance = 0.735 0.903 0.903 0.903 obstacle-direction = left`

Fig. 7.12. Optical flow interpretations of highlighted waypoint images in Fig. 7.11

7.3. Experiment of motion control with path planning EXP-MP

The objective of the following experiments, called EXP-MP, was to verify the operation of the behavior-based control system in complicated situations, concerning both motion control and topological path planning. The software structure for experiments EXP-MP is illustrated in Fig. 7.13.

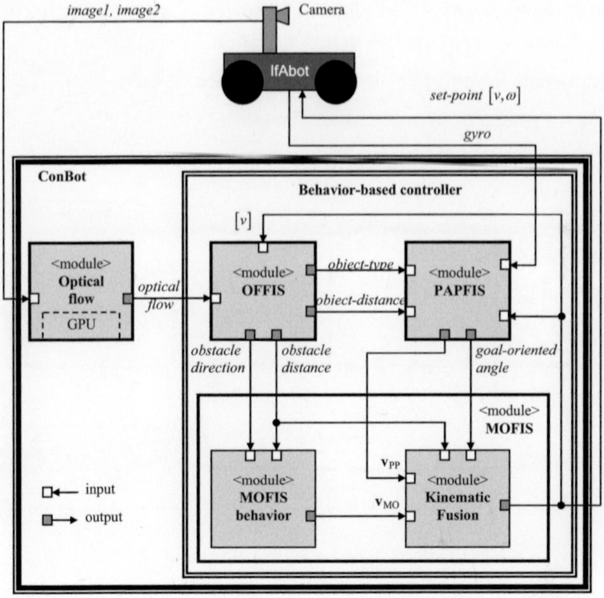

Fig. 7.13. Software structure of the navigation system for the experiments

7.3.1. Topological map

The experiments EXP-MP were concerned with high-level navigation tasks in the office, where inherent landmarks such as walls, corners, doors, and corridors can be used as positional cues for topological path planning.

Fig. 7.14 displays the 12 indoor landmarks that were selected to create the topological map of the experimental environment, as illustrated in Fig. 7.15. Additionally, the four basic directions of the experimental environment {N, E, S, W} measured by the gyroscope, shown in Fig. 7.15, were necessary for topological path planning.

The node and edge attributes of the topological map are defined by the table form shown in Fig. 7.16. Most of the edges are a two-way route, e.g. existence of E_{2_3} and E_{3_2}. However, some of them are one-way routes, e.g. existence of E_{7_9} but inexistence of E_{9_7}.

The node attributes compose the node ID with symbols N_1 to N_{12}, illustrated in the first column of the table, and a landmark feature *LmkF*, illustrated in the second column of the table. A landmark feature *LmkF* includes a landmark type *Lmk* and a landmark depth Dp_G.

As explained in chapter 6, Dp_G is the approximate depth of a hollow goal, e.g. the Dp_G of a door is a relative distance from one side to other side of the door and the Dp_G of a corridor is an approximate distance of the real corridor length that should be bigger than half of this length and smaller than this length.

Notice that the Dp_G of walls or corners is always zero because mobile robots can never go through walls and corners.

The edge attributes comprise the edge ID, illustrated in the third column of the table, with symbols $E_{i\,j}$, where $i, j = 1 \ldots 12$, and an action feature *ActF*, illustrated in the last column of the table. An action feature *ActF* contains a goal direction θ_G, $\theta_G \in \{N, E, S, W\}$, and a motion action at a goal act_G, $act_G \in \{Goto, Though\}$.

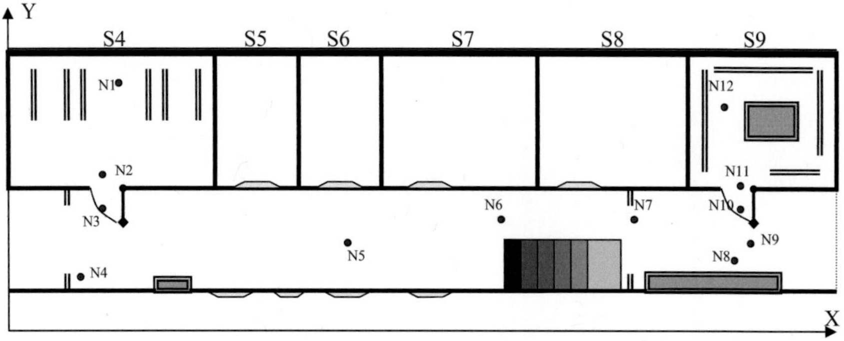

Fig. 7.14. Inherent landmarks selected in experiments EXP-MP

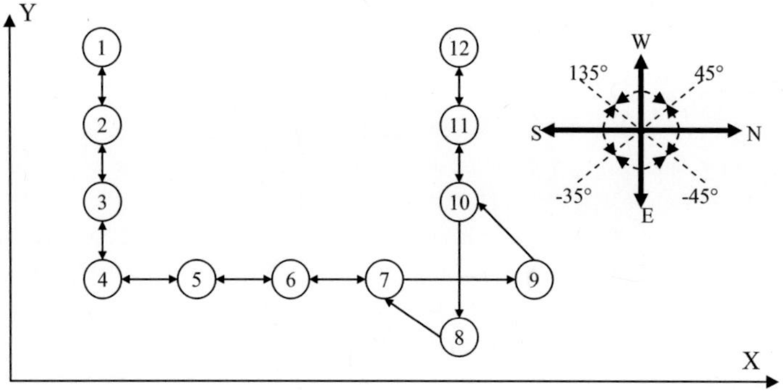

Fig. 7.15. Topological map and four basic directions of the gyroscope in experiments EXP-MP

Node ID	Landmark feature *LmkF*	Edge ID	Action feature *ActF*
N_1	*Lmk* = *Wall* Dp_G = 0 cm	E_{1_2}	θ_G=E, act_G= *Goto*
N_2	*Lmk* = *Door* $Dp_G \approx$ 200 cm	E_{2_1}	θ_G =W, act_G = *Goto*
		E_{2_3}	θ_G =E, act_G = *Through*
N_3	*Lmk* = *Door* $Dp_G \approx$ 200 cm	E_{3_2}	θ_G =W, act_G = *Through*
		E_{3_4}	θ_G =E, act_G = *Goto*
N_4	*Lmk* = *Corner* $Dp_G \approx$ 0 cm	E_{4_3}	θ_G =W, act_G = *Goto*
		E_{4_5}	θ_G =N, act_G = *Through*
N_5	*Lmk* = *Corridor* $Dp_G \approx$ 800 cm	E_{5_4}	θ_G =S, act_G = *Goto*
		E_{5_6}	θ_G =N, act_G = *Goto*
N_6	*Lmk* = *Door* Dp_G = 400 cm	E_{6_5}	θ_G =S, act_G = *Through*
		E_{6_7}	θ_G =N, act_G = *Through*
N_7	*Lmk* = *Door* $Dp_G \approx$ 400 cm	E_{7_6}	θ_G =S, act_G = *Through*
		E_{7_9}	θ_G =N, act_G = *Goto*
N_8	*Lmk* = *Wall* Dp_G = 0 cm	E_{8_7}	θ_G =S, act_G = *Goto*
N_9	*Lmk* = *Door* $Dp_G \approx$ 200 cm	E_{9_10}	θ_G =W, act_G = *Goto*
N_{10}	*Lmk* = *Door* $Dp_G \approx$ 200 cm	E_{10_8}	θ_G =E, act_G = *Goto*
		E_{10_11}	θ_G =W, act_G = *Through*
N_{11}	*Lmk* = *Door* $Dp_G \approx$ 200 cm	E_{11_10}	θ_G =E, act_G = *Through*
		E_{11_12}	θ_G =W, act_G = *Goto*
N_{12}	*Lmk* = *Wall* $Dp_G \approx$ 0 cm	E_{12_11}	θ_G =E, act_G = *Goto*

Fig. 7.16. Node and edge attributes of the topological map for experiments EXP-MP

7.3.2. Experiment with inherent obstacles EXP-MP-i

In the experiments EXP-MP-i, the navigation tasks of the mobile robot were to travel from a certain node to another node given by the user, and simultaneously avoid collisions from the inherent obstacles, i.e. the inherent furniture of the office.

In experiment EXP-MP-i1, the robot started from node N_1 and travelled to node N_5. Based on the graph-based search algorithm described in chapter 5, the behavior-based controller automatically found the shortest way $\{N_1, N_2, N_3, N_4, N_5\}$ to the target. Then it controlled the robot while it followed its path.

The driven trajectory of the IfAbot is illustrated by the ground truth in Fig. 7.17 with several highlighted waypoint images ImE31, ImE32 and ImE33.

Fig. 7.18 displays the optical flow interpretations of the OFFIS module with respect to the highlighted situations in Fig. 7.17.

In experiment EXP-MP2, the IfAbot executed the navigation task of moving from node N_{12} to node N_6.

The global path $\{N_{12}, N_{11}, N_{10}, N_8, N_7, N_6\}$ was automatically found by the path planner in the PAPFIS module.

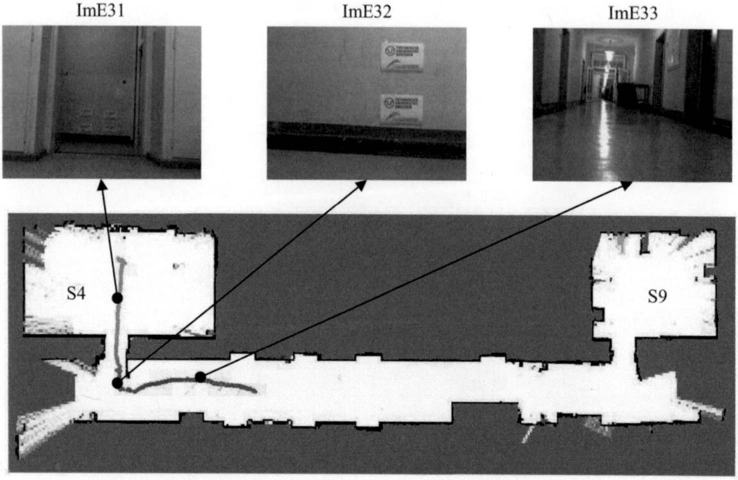

Fig. 7.17. Ground truth of the driven trajectory of the IfAbot in experiment EXP-MP-i1

Optical flow images (Hierarchical segmentation)	**Optical flow interpretations (OFFIS outputs)**

ImE31
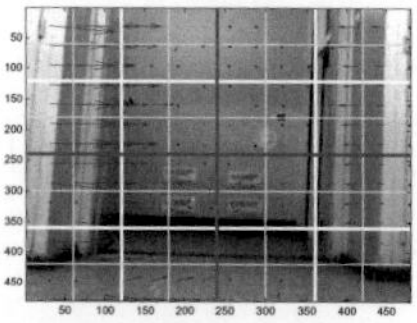
	`obstacle-distance =`
	`0.517 0.903 0.903 0.903`
	`obstacle-direction = left`
	
	`TPO-distance =`
	`0.517 0.903 0.903 0.903`
	`TPO-type = door`

ImE32
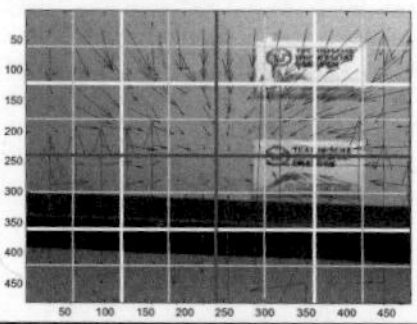
	`obstacle-distance =`
	`0.610 0.620 0.310 0.429`
	`obstacle-direction = right`
	
	`TPO-distance =`
	`0.698 0.620 0.586 0.714`
	`TPO-type = wall`

ImE33
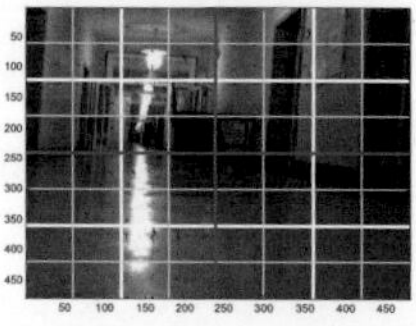
	`obstacle-distance =`
	`0.903 0.806 0.704 0.874`
	`obstacle-direction = right`
	
	`TPO-distance =`
	`0.903 0.903 0.903 0.903`
	`TPO-type = corridor`

Fig. 7.18. Optical flow interpretations of highlighted waypoint images in Fig. 7.17

The result of experiment EXP-MP-i2 is illustrated in Fig. 7.19 by the ground truth of the driven trajectory of the IfAbot with some highlighted waypoint images ImE41, ImE42, ImE43 and ImE44. The optical flow interpretations of the OFFIS module with respect to the highlighted waypoint images are shown in Fig. 7.20.

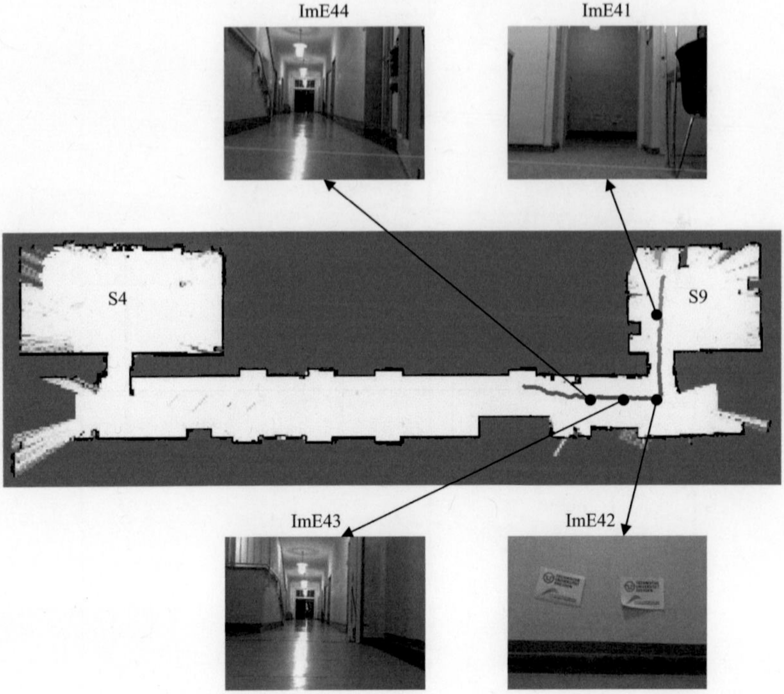

Fig. 7.19. Ground truth of the driven trajectory of the IfAbot in experiment EXP-MP-i2

Optical flow images (Hierarchical segmentation)	**Optical flow interpretations (OFFIS outputs)**
ImE41 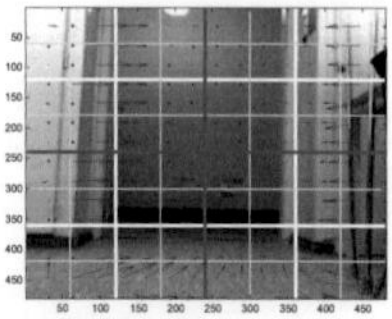	*obstacle-distance =* *0.734 0.903 0.903 0.897* *obstacle-direction = left* *TPO-distance =* *0.782 0.903 0.903 0.899* *TPO-type = door*
ImE42 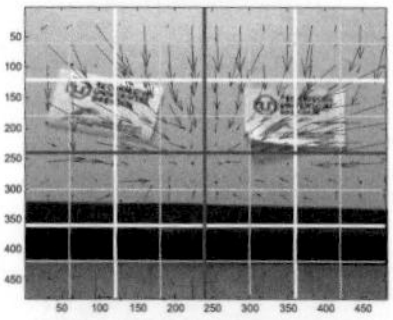	*obstacle-distance =* *0.637 0.565 0.893 0.553* *obstacle-direction = right* *TPO-distance =* *0.721 0.565 0.598 0.553* *TPO-type = wall*
ImE43 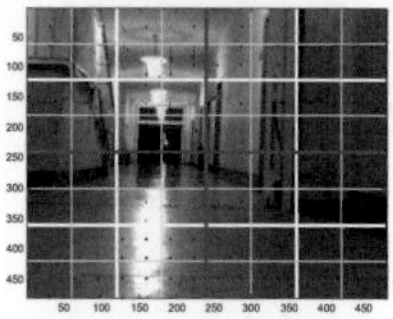	*obstacle-distance =* *0.903 0.903 0.903 0.903* *obstacle-direction = none* *TPO-distance =* *0.903 0.903 0.903 0.903* *TPO-type = door*
ImE44 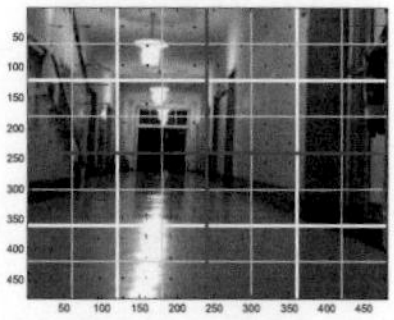	*obstacle-distance =* *0.903 0.903 0.903 0.903* *obstacle-direction = none* *TPO-distance =* *0.903 0.903 0.903 0.903* *TPO-type = corridor*

Fig. 7.20. Optical flow interpretations of highlighted waypoint images in Fig. 7.19

7.3.3. Experiment with artificial obstacles EXP-MP-a

Like the experiments EXP-MP-i, in the experiments EXP-MP-a the IfAbot executed the navigation tasks of travelling from a certain node to another node given by the user, and avoiding collisions from the inherent furniture of the office.

Unlike the experiments EXP-MP-i, in the experiments EXP-MP-a the obstacle avoidance task of the IfAbot was concerned in avoiding artificial obstacles randomly strewn along corridor crd49. Additionally, the positions of these artificial obstacles were changed from those in experiments EXP-MO.

In experiment EXP-MP-a1, the navigation task of the robot was to travel from node N_1 (in room S4) to node N_7 (the corridor door in corridor crd49). The global path to the target $\{N_1, N_2, N_3, N_4, N_5, N_6, N_7\}$ was automatically identified by the PAPFIS module.

The result of the experiment EXP-MP-a1 is shown in Fig. 7.21 by the ground truth of the driven trajectory of the IfAbot with several highlighted waypoint images ImE51, ImE52, ImE53, ImE54 and ImE55. Then, Fig. 7.22 illustrates the optical flow interpretations of the OFFIS module with respect to the highlighted situations displayed Fig. 7.21.

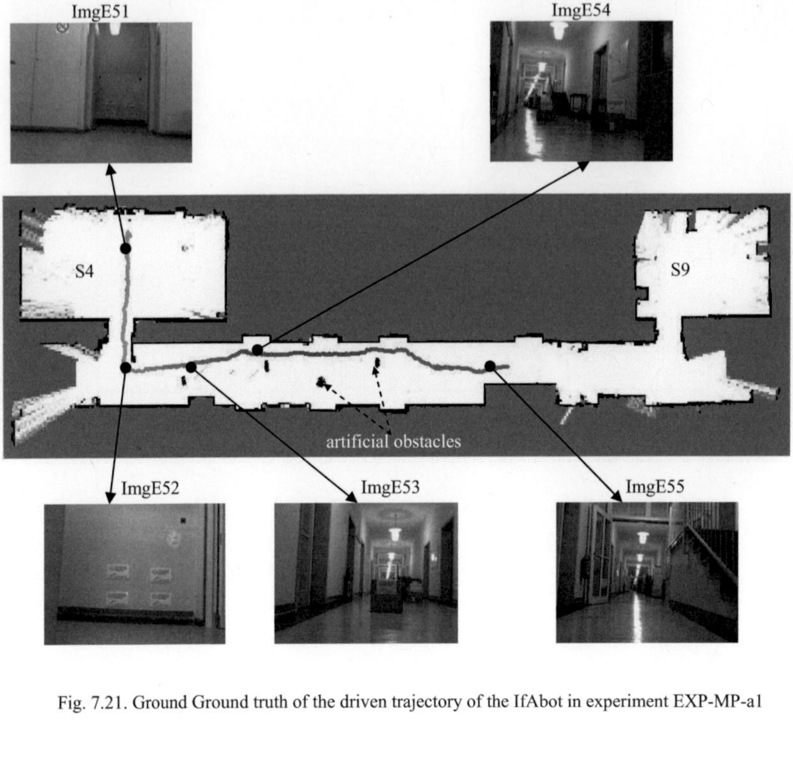

Fig. 7.21. Ground Ground truth of the driven trajectory of the IfAbot in experiment EXP-MP-a1

Optical flow images (Hierarchical segmentation)	Optical flow interpretations (OFFIS outputs)
ImE51 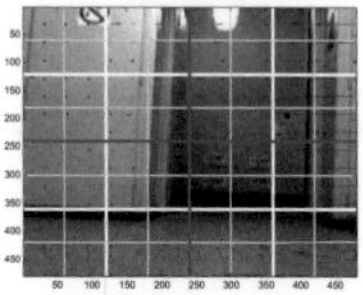	*obstacle-distance = 0.903 0.903 0.903 0.845 obstacle-direction = none TPO-distance = 0.903 0.903 0.903 0.845 TPO-type = door*
ImE52 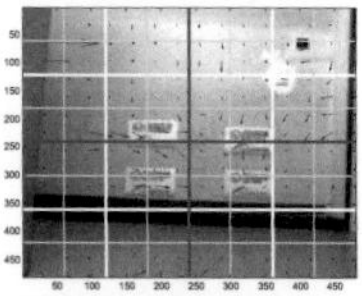	*obstacle-distance = 0.903 0.903 0.903 0.903 obstacle-direction = none TPO-distance = 0.903 0.903 0.903 0.903 TPO-type = corner*
ImE53 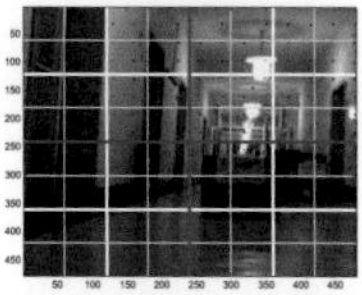	*obstacle-distance = 0.903 0.903 0.903 0.903 obstacle-direction = none TPO-distance = 0.903 0.903 0.903 0.903 TPO-type = corridor*
ImE54 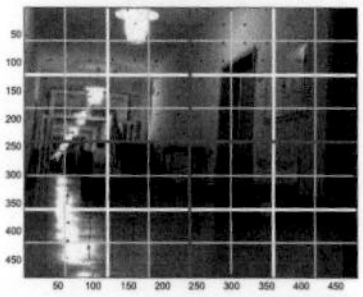	*obstacle-distance = 0.903 0.903 0.903 0.663 obstacle-direction = right TPO-distance = 0.903 0.903 0.903 0.903 TPO-type = corridor*
ImE55 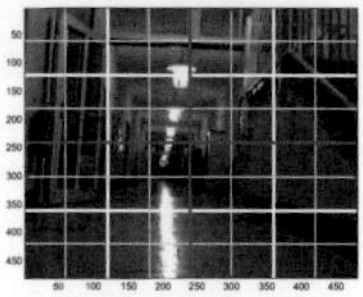	*obstacle-distance = 0.903 0.903 0.903 0.886 obstacle-direction = none TPO-distance = 0.903 0.903 0.903 0.886 TPO-type = door*

Fig. 7.22. Optical flow interpretations of highlighted waypoint images in Fig. 7.21

In experiment EXP-MP-a2, the artificial obstacles were randomly arranged from the inside of room S9 to corridor crd49. Also, an artificial obstacle was purposefully placed in front of door dr9 of room S9 as seen in the illustration in Fig. 7.23.

The navigation task of the IfAbot comprises travelling from node N_{12} (in room S9) to node N_5 (in the middle of corridor crd49) and avoiding the obstacles. The global planner automatically found the shortest way to the target, which was $\{N_{12}, N_{11}, N_{10}, N_8, N_7, N_6, N_5\}$.

The results of experiment EXP-MP-a2 is illustrated in Fig. 7.24 by the ground truth of the driven trajectory of the IfAbot with some highlighted images ImE61, ImE62, ImE63 and ImE64. The optical flow interpretations of these highlighted images are listed in Fig. 7.25.

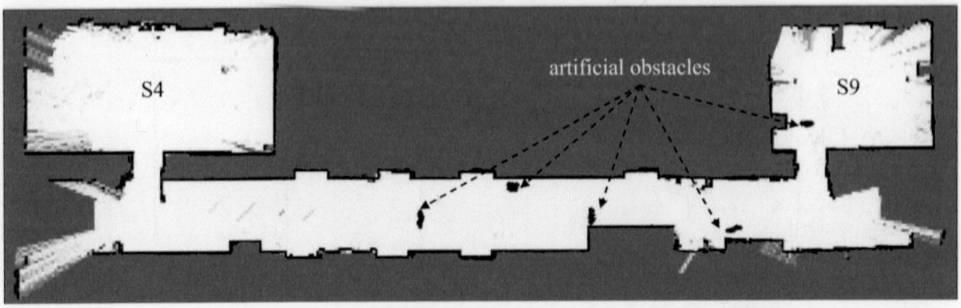

Fig. 7.23. Laser-scanned plan of experiment EXP-MP-a2

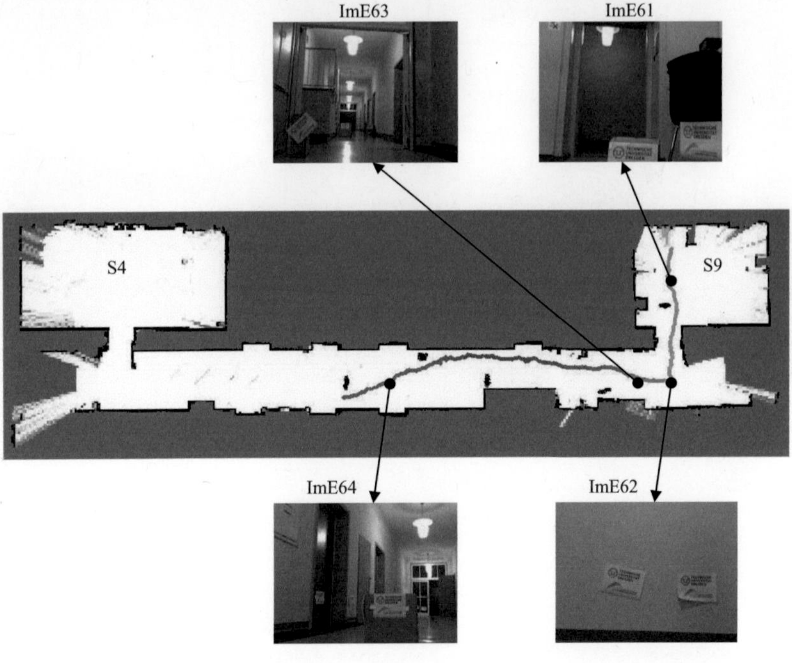

Fig. 7.24. Ground truth of the driven trajectory of the IfAbot in experiment EXP-MP-a2

Optical flow images (Hierarchical segmentation)	Optical flow interpretations (OFFIS outputs)
ImE61 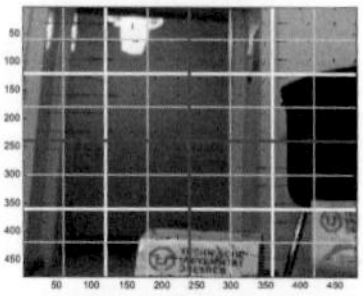	*obstacle-distance =* *0.903 0.903 0.903 0.703* *obstacle-direction = right* *TPO-distance =* *0.718 0.903 0.903 0.903* *TPO-type = door*
ImE62 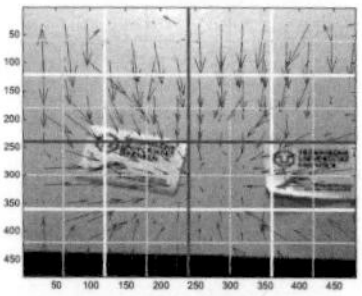	*obstacle-distance =* *0.618 0.683 0.488 0.683* *obstacle-direction = right* *TPO-distance =* *0.618 0.683 0.488 0.705* *TPO-type = wall*
ImE63 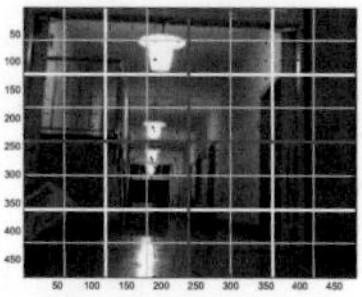	*obstacle-distance =* *0.903 0.903 0.903 0.903* *obstacle-direction = none* *TPO-distance =* *0.903 0.903 0.903 0.903* *TPO-type = door*
ImE64 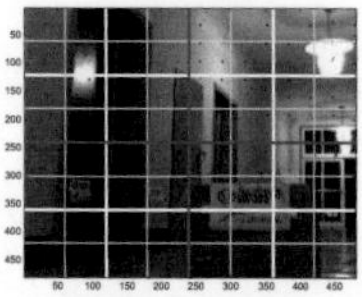	*obstacle-distance =* *0.903 0.903 0.461 0.903* *obstacle-direction = right* *TPO-distance =* *0.903 0.903 0.903 0.903* *TPO-type = corridor*

Fig. 7.25. Optical flow interpretations of highlighted waypoint images in Fig. 7.24

7.4. Computation time of functional modules

The computation times of the different modules of the behavior-based controller are displayed in Fig 7.26. Fig. 7.26 shows that most of the computation time of the controller is spent on the optical flow-based perception at the OFFIS module.

Modules of the behavior-based controller	Computation time (ms)
OFFIS module	75.27
PAPFIS module	11.31
MOFIS module	18.88

Fig. 7.26. Computation time of the modules of the behavior-based controller

Sub-modules of the OFFIS module	Computation time (ms)
Hierarchical segmentation	1.70
Outlier detection filter	1.21
Quadrant averaging	9.20
Structural identification	1.05
Fuzzy distance calculation	10.89
Fuzzy-based interpretation	51.22

Fig. 7.27. Computation time of the sub-modules of the OFFIS module

Functions of PAPFIS module	Computation time (ms)
Global and local planner	0.28
Primitive behaviors	5.48
Composite behaviors	5.50
Behavior selector	0.05

Fig. 7.28. Computation time of the sub-modules of the PAPFIS module

Functions of MOFIS module	Computation time (ms)
MOFIS behaviors	7.24
MOFIS behavior selector	0.05
Fuzzy-based kinematic fusion	11.59

Fig. 7.29. Computation time of the sub-modules of the MOFIS module

For more details, the computation times of the sub-modules of the OFFIS, PAPFIS and MOFIS modules are respectively listed in Fig. 7.27, 7.28, and 7.29. They prove that the computation time of the sub-modules that apply fuzzy inference systems takes much more computation time than that of other sub-modules.

7.5. Discussion

On one hand, the results of experiments EXP-MO prove the reliable capability of the designed control system for low level navigation tasks concerning obstacle avoidance in risky environments. Noticeably, the applications of the fuzzy-based behaviors to the collision-free navigation made the mobile robot motion flexible and reliable in unknown and complex environments.

On the other hand, the results of experiments EXP-MP demonstrate the robust capability of the behavior-based control system for the high-level navigation tasks concerning topological path planning in complex environments. Especially, the robot does not need exact coordinates for its positions. This means the problem of accumulated error position is no longer an issue. This allows for the simple operation of the control system.

Moreover, the experimental results show the successful use and the reliable operation of topological optical flow perception using fuzzy logic in the real environment, and also prove that the outlier detection filter improves significantly the capability of tolerating ambiguities of the optical flow perception.

Furthermore, for the computation time values it should be recognized that the current implementation uses Matlab m-files and the standard Matlab Fuzzy-Toolbox and is thus not optimized at all with respect to computation time. The absolute computation time can be reduced significantly by using dedicated embedded software implementations like C/C++.

Generally speaking, the experimental results prove that the modular architecture of the behavior-based control system is generally applicable for the mobile robots concerning behavior-based control and topological path planning. They also prove that the optical flow perception is reliable in real environments under all possible noise effects, e.g. unfavorable lightning conditions, non-standard objects, image processing limitations, image noise, etc.

Chapter 8
Discussion and conclusion

This chapter concludes the work presented throughout this thesis. Firstly, it discusses the theoretical and experimental results of the works. Then, it indicates the contributions of the thesis. Finally, the outlook is described.

8.1. Discussion

Compared to the classic metric approaches, the proposed approach for optical flow-based perception uses *much simpler arithmetic* and consumes *less computation time* because of the use of *qualitative* optical flow interpretation.

This approach is particularly suitable for equally-spaced and homogeneously-distributed, but not necessarily highly-dense, optical flow fields, such as those generated by fast analog or hybrid optical flow sensors.

The performances of the qualitative optical flow perception using fuzzy logic were successfully tested in experiments within a real office environment. Stemming from these experimental results, it has been proven that the optical flow-based perception is reliable in real environments including relevant noise effects, e.g. unfavorable lightning conditions, non-standard objects, image processing limitations, image noise, etc.

Additionally, the optical flow perception supports information about obstacle directions in a 2D matrix, including horizontal and vertical directions, without any metric information. This enables obstacle avoidance, which is applicable for visual mobile robots working not only on flat surfaces, but also in different terrains, e.g. under water or in space.

Furthermore, the algorithms of the optical flow perception enable the interpretation quality to be improved by using deeper layers for the Pyramid method and a simple development of a bigger number (>4) for distinguishable objects by adding sub-modules into the OFFIS rough and fine classifiers.

On the other hand, the new generic modular hierarchical architecture for a behavior-based control system with clear interfaces to path planning and motion control systems allows new navigation tasks to be easily adapted by adding or changing behavior modules. Additionally, the fuzzy logic concepts applied to the behaviors, which are incorporated in the hierarchical architecture, make robot behavior more flexible and robust with respect to changing environments (dynamic obstacles) and following biological paradigms.

The successful results of the experiments in a real environment prove the reliability and robustness of the behavior-based control system integrated with the qualitative optical flow perception working in real environments.

Moreover, the simulation and experiment results also prove that the combination of topological path planning and fuzzy-based behaviors organized in a hierarchical structure allows mobile robots to perform high-level navigation tasks in complex and changing environments though lacking metric information.

Generally speaking, the optical flow perception and the modular architecture of the behavior-based control system are able to be used for visual mobile robots working in real environments under relevant noise effects.

8.2. Contributions

To start off, this work has presented a certain type of bio-inspired optical flow processing in terms of a qualitative topological optical flow perception for visual mobile robots in order to obtain structural information of the working environment. Compared to the classic metric approaches, this approach uses much simpler arithmetic and consumes less computation time because of the use of qualitative optical flow interpretation. Additionally, it provides mobile robots with visual perception that enables more 3D detail about the working environment to be obtained.

Then a modular architecture was presented for behavior-based mobile robot control systems using fuzzy logic. The behaviors are functionally classified into two separate modules, concerning collision-free motion control and goal-oriented path planning using topological maps. The applied fuzzy logic approach allows for flexible and robust behavior design in complex environments.

Finally, a new scheme for topological path planning in combination with fuzzy-based behaviors has been introduced for the goal-oriented navigation of a mobile robot. The designed scheme allows a mobile robot to perform topological path planning in a real environment without the metric information from the coordinates and positions of the mobile robot and enables an easy adjustment of topological path planning for different sensor perceptions or landmarks by just changing the topological map data.

In comparison with the objectives stated in chapter 1, the achieved results of this work are listed as follows:

- ❖ **Optical flow-based perception**. A new approach for optical flow-based perception through qualitative interpretations to obtain structural information from the working environment has been developed. Compared to the conventional approaches for optical flow-based perception, this approach has the advantage of supplying simple and effective methodologies for processing optical flow in qualitative manners. This approach enables the optical flow interpretations to use much simpler arithmetic and consume less computation time. Besides, by using fuzzy logic, the optical flow perception becomes bio-inspired optical flow perception, which enables a robust perception despite the noisy conditions of the working environment.
- ❖ **Generic modular structure of behavior-based control system**. A generic modular structure of behavior-based control system has been implemented with three distinct modules for perception, path planning and motion control. These modules are connected by simple IO interfaces. The system is able to be used for mobile robots by just changing the sensor-based perception type. The hierarchical organization of behaviors makes system operation more efficient.
- ❖ **Fuzzy-based behaviors and fusion**. The application of fuzzy logic for behaviors allows flexible and robust behavior design in complex environments. Additionally, the

fuzzy-based fusion of behaviors helps mobile robots solve conflicts among the behaviors more rationally.

❖ **Topological path planning**. A generalized scheme for topological path planning combined with fuzzy-based behaviors organized in a hierarchical architecture to navigate a mobile robot to a given target has been performed. This scheme helps mobile robots execute high-level navigation tasks in scenarios that lack metric information about the working environments.

8.3. Outlook

After the objectives of this work have been achieved, there are several open issues, which remain for the future research. The main open issues can be summarized as follows:

- Development of an algorithm of optical flow-based interpretations for a larger number of distinguishable indoor landmarks, i.e. more than four TPO types.
- Development of optical flow-based interpretations for collision-free navigation of mobile robots working in outdoor environments, e.g. under water or in space.
- Improvement of the quality of optical flow-based interpretation by applying deeper layers of the hierarchical segmentation, i.e. the Pyramid approach.
- Enlargement of the scope of the optical flow-based perception by using other cameras with a larger field of view, e.g. omni-camera.

References

Abraham, A. T., S. S. Ge and P. Y. Tao. (2009). "A topological approach of path planning for autonomous robot navigation in dynamic environments," In International Conference on Intelligent Robots and Systems, IROS, pp. 4907 – 4912.

Aires, K. R. T. and A. A. D. de Medeiros. (2010). "A Plane Segmentation System Based on Affine Homography and Optical Flow," Conference on Graphics, Patterns and Images (SIBGRAPI), pp. 346-352.

Andersen, N. A., L. Henriksen, and O. Ravn. (1995). "Visual Positioning and Docking of Non-holonomic vehicles," in Proc. 4th Int. Symp. Experimental Robot, vol. 223, pp. 140-149.

Arkin, R. C. (1999). *Behavior-based Robotics*, Cambridge, Mass.

Bao, Q., S. Li, W. Shang and M. An. (2009). "A Fuzzy Behavior-Based Architecture for Mobile Robot Navigation in unknown environments," Int. Conf. on Artificial Intelligence and Computational Intelligence, AICI '09, pp. 257 - 261.

Barrows, G. and C. Neely. (2000). "Mixed-mode VLSI optic flow sensors for in-flight control of a micro air vehicle ", Critical Technol. Future Comput., SPIE, vol. 4109, pp. 52-63.

Beauchemin, S. and J. L. Barron. (1995). "The computation of optical flow", ACM Computing Surveys (CSUR), vol. 27, no. 3, Sep. 1995, pp. 433 – 466.

Beck, M. (2004). *Ein Navigationsverfahren für mobile Roboter unter Nutzung des Optischen Flusses*, TU Dresden, Dresden.

Beom, H. R. and H.S. Cho. (1995). "A sensor-based navigation for a mobile robot using fuzzy Logic and Reinforcement Learning", IEEE Transactions on Systems, Man and Cybernetics, vol. 25, pp. 464-477.

Berns, K., and E. v. Puttkamer. (2009). *Autonomous Land Vehicles*, Vieweg + Teubner.

Brooks, R. (1986). "A robust layered control system for a mobile robot," IEEE Journal of Robotics and Automation, vol. 2, issue 1, pp. 14-23.

Carreira, F., T. Canas, A. Silva and C. Cardeira. (2006). "i-Merc: A mobile robot to deliver meals inside health services," IEEE Conference on Robotics, Automation and Mechatronics, pp. 1-8.

Carreira, F., T. Canas, A. Silva and C. Cardeira. (2007). "A fuzzy controller for a health services mobile robot," IEEE International Symposium on Industrial Electronics, pp. 3287-3292.

Chai, X., F. Wen and K. Yuan. (2011). "Fast Vision-based Object Segmentation for natural landmark detection on indoor mobile robot," Proceedings of the 2011 IEEE Int. Conf. on Mechatronics and Automation, Beijing, China, pp. 2232-2237.

Chen, Q., Z. Sun, D. Liu, L. Cao and T. Chen. (2011). "Integrating topological map for autonomous navigation in unstructured environments," in International Conference on Intelligent Control and Information Processing (ICICIP), pp. 185 –189.

Chiang, C. H., P. J. Chiang, J. C. C. Fei and J. S. Liu. (2007). "A comparative study of implementing Fast Marching Method and A* search for mobile robot path planning in grid environment: Effect of map resolution," Proc. IEEE Workshop Adv. Rob. Soc. Impacts, ARSO, pp. 1-6.

Coombs, D. and K. Roberts. (1993). "Centering behavior using peripheral vision," in Proc. of IEEE Conf. on Computer Vision and Pattern Recognition, IEEE Computer Society Press, Los Alamitos, CA, pp. 440–451.

Dellaert, F., D. Fox, W. Burgard, and S. Thrun. (1999). "Monte Carlo Localization for Mobile Robots," Proc. IEEE Int. Conf. Robotics and Automation, pp. 1322-1328

Dev, A., B. Kröse and F. Groen. (1997). "Navigation of a Mobile Robot on the Temporal Development of the Optic Flow," in Proc. IEEE Int. Conf. Intelligent Robots and Systems, pp. 558-563.

Díaz, J., E. Ros, F. Pelayo, E. M. Ortigosa and S. Mota. (2006). "FPGA-Based Real-Time Optical-Flow System", IEEE Transactions on Circuits and Systems for Video Technology, vol. 16, no. 2, pp. 274-279.

ElKaissi, M., M. Elgamel, M. Bayoumi and B. Zavidovique. (2007). "SEDLRF: A New Door Detection System for Topological Maps," in International Workshop on Computer Architecture for Machine Perception and Sensing, pp. 75 –80.

El-Teleity, S. A. L., Z. B. Nossair, H. M. A. K. Mansour and A. TagElDein. (2011). "Fuzzy logic control of an autonomous mobile robot," in the 16th International Conference on Methods and Models in Automation and Robotics (MMAR), pp. 188-193.

Enkelmann, W. (1991). "Obstacle detection by evaluation of optical flow fields from image sequences," Image Vision Comput. no. 9, pp.160-168.

Evans, J. (1994). "HelpMate: An autonomous mobile robot courier for hospitals," Int. Conference on Intelligent Robots and Systems, IROS'94, vol. 3, pp. 1695-1700

Florczyk, S. (2005). *Robot Vision*, Wiley-VCH, pp. 36-45.

Fox, D., W. Burgard, and S, Thrun. (1997). "The dynamic window approach to collision avoidance," Robotics & Automation Magazine IEEE, vol. 4, iss. 1, pp. 23–33.

Fraichard, T. (1993). "Dynamic trajectory planning with dynamic constraints: A `state-time space' approach," IEEE/RSJ International Conference on Intelligent Robots and Systems, vol. 2, pp. 1393-1400.

Franz, M. O. and H. A. Mallot. (2000). "Biomimetic robot navigation," Robotics and Autonomous Systems 30, pp. 133-153.

Harris, C. and M. Stephens. (1988). "A Combined Corner and Edge Detector," Plessey Research Roke Manor, United Kingdom, Plessey Company plc, pp. 147-151.

Horn, B. K. P. and B. G. Schunck. (1981). "Determining Optical Flow", Artificial Intelligence, vol.17, pp.185-204.

Iu, S. L., und Lin, Y. T. (1999). "Re-examining the optical flow constraint. A new optical flow algorithm with outlier rejection," in International Conference on Image Processing, ICIP 99. Proceedings, vol.3, pp. 727–731.

Janschek, , K., V. Tchernykh and M. Beck. (2006). "Performance Analysis for Visual Planetary Landing Navigation using Optical Flow and DEM Matching," in: Proc. of the AIAA Guidance, Navigation and Control Conf., Aug. Paper no. AIAA-2006-6706, Keystone, CO.

Jeong, W. Y. and K. M. Lee. (2006). "Visual SLAM with Line and Corner Features," in Proc. IEEE/RSJ Int. Conf. Intell. Robots Syst., Beijing, China, pp. 2570-2575.

Jia, Q. and X. Wang. (2009). "Path planning for mobile robots based on a modified potential model," IEEE International Conference on Mechatronics and Automation ICMA, pp. 4946-4951.

Khatib, O. (1985). "Real-time obstacle avoidance for manipulators and mobile robots," in Proceedings of the IEEE Int. Conference on Robotics and Automation, vol. 2, St. Louis, MO, pp. 500-505.

Kimmer, S. (2009). *Utilization of Graphics Processing Units (GPU) for fast computing of the optical flow*, Studienarbeit, TU Dresden, Dresden.

Kittler, J. and F. M. Alkoot. (2003). "Sum versus vote fusion in multiple classifier systems," IEEE Transactions on Pattern Analysis and Machine Intelligence, vol. 25, no. 1, pp. 110-115.

Krishnamurthy, B. and J. Evans. (1992). "HelpMate: A robotic courier for hospital use," in IEEE International Conference on Systems, Man and Cybernetics, vol. 2, pp. 1630-1634.

Lee, K., N. Cho, W. K. Chung and N. L. Doh. (2006). "Topological Navigation of Mobile Robot in Corridor Environment using Sonar Sensor," in International Conference on Intelligent Robots and Systems, pp. 2760 –2765.

Lei, Y. (2010). "The behavior fusion based on Particle Swarm Optimization method," International Conference on Computer, Mechatronics, Control and Electronic Engineering (CMCE), Changchun, China, pp. 380-383.

Lin, W., S. Jia, F. Yang and K. Takase. (2004). "Topological navigation of mobile robot using ID tag and WEB camera," in International Conference on Intelligent Mechatronics and Automation, pp. 644 – 649.

Liu, H., T. H. Hong, M. Herman, T. Camus and R. Chellappa. (1998). "Accuracy vs Efficiency Trade-offs in Optical Flow Algorithms", Computer Vision and Image Understanding, vol. 72, no. 3, pp. 271-286.

Mai, N. A. and K. Janschek. (2009). "Bio-inspired Optical Flow Interpretation with Fuzzy Logic for Behavior-Based Robot Control," In: Proc. of the RAAD 2009 18th International Workshop on Robotics in Alpe-Adria-Danube Region, Brasov, Romania, pp. 68-77.

Mayes, P. (1990). "Situated agents can have goals," Robotics and Automation Systems 6, pp. 49-70.

MB385, (BlueBotics, www.bluebotics.com/automation/MB835/).

Mechlih, H. (1993). "Trajectory planning for mobile robots in a dynamic environment," IEEE-IEE Conference on Vehicle Navigation and Information Systems, pp. 551-554.

Michel, P., J. Chestnutt, J. Kuffner and T. Kanade. (2005). "Vision-guided humanoid footstep planning for dynamic environments," International Conference on Humanoid Robots, 5th IEEE-RAS, Japan, pp. 13-18.

Modes, V. (2010). *Inbetriebnahme eines radgetriebenen mobilen Demonstrators sowie Realisierung von Navigationsaufgaben*, Studienarbeit, TU Dresden, Dresden.

Naito, T., T. Ito and Y. Kaneda. (2007). "The Obstacle Detection Method using Optical Flow Estimation at the Edge Image," Intelligent Vehicles Symposium, 2007 IEEE, pp. 817 – 822.

Nalbach, G. and R. Hengstenberg. (1994). "The halteres of the blowfly calliphora - Three-dimensional organization of compensatory reactions to real and simulated rotations", J. Comparative Physiol. A, vol. 175, pp. 695-708.

Ohnishi, N. and A. Imiya. (2006). "Dominant plane detection from optical flow for robot navigation," Pattern Recognition Letters, vol. 27, no. 9, pp. 1009-1021.

Pauly, M., M. Finke, L. Peters and K. Beck. (1999). "Control and service structure of a robot team," International Conference on Intelligent Robots and Systems, IROS '99, vol. 2, pp. 1069-1074.

Pfeifer, R., F. Iida and J. Bongard. (2005). "New Robotics: Design Principles for Intelligent Systems." Journal Artificial Life, volume 11, issue 1-2, pp. 99–120.

Poggio, T., A. Verri, and V. Torre. (1991). *Green theorems and qualitative properties of the optical flow*, Mass. Inst. Technol., Cambridge, MA, Tech. Rep. A.I. Memo 1289.

Pollack, M. E., L. Brown, D. Colbry, C. Orosz, B. Peintner, and S. Ramakrishnan. (2002). "Pearl: A mobile robotic assistant for the elderly," in AAAI Workshop on Automation as Caregiver.

POV-Team. (Stand:15.12.2011). URL: http://www.povray.org/documentation/

Prazdny, K. (1980). "Egomotion and relative depth map from optical flow," Biological Cybernetics (Historical Archive), vol. 36, issue 2, pp. 87-102.

Russell, S. and P. Norvig. (2003). *Artificial Intelligence - A modern approach*, Prentice Hall.

Saffiotti, A., K. Konolige and H. E. Ruspini. (1995). "A multivalued-logic approach to integrating planning and control," Artificial Intelligence 76, pp. 481-526.

Santos-Victor, J., G. Sandini, F. Curotto and S. Garibaldi. (1993). "Divergent stereo for robot navigation: Learning from bees," in Proc. IEEE CS Conf. Computer Vision and Pattern Recognition, pp. 434-439.

Santos-Victor, J., G. Sandini, F. Curotto and S. Garibaldi. (1995). "Divergent stereo for robot navigation: A step forward to a robotic bee", Intl. Journal of Computer Vision 14, pp. 159-177

Schnellhorn, F. (2009). *Entwurf eines radgetriebenen mobilen Demonstrators für Navigationsaufgaben in Büroumgebungen bzw. in büronahen Außenbereichen*, Studienarbeit, TU Dresden, Dresden.

Sedigh, K. H., K. Ashenayi, T. W. Manikas, R. L. Wainwright and H. M. Tai. (2004). "Autonomous Local Path-Planning for a Mobile Robot using a genetic algorithm," Congress on Evolutionary Computation CEC2004, vol. 2, pp. 1338-1345.

Seraji, H. and A. Howard. (2002). "Behavior-Based Robot Navigation on Challenging Terrain: A Fuzzy Logic Approach," IEEE Transactions on Robotics and Automation, vol. 18, pp. 308-321.

Shibata, M., T. Makino and M. Ito. (2008). “Target distance measurement based on camera moving direction estimated with optical flow,” in 10th IEEE International Workshop Advanced Motion Control, pp. 62-67.

Simmons, R. and S. Koenig. (1995). “Probabilistic Robot Navigation in Partially Observable Enironments,” Proc. Int. Joint Conf. Artificial Intelligence, pp. 1080-1087.

Srinivasan, M. V., M. Lehrer, W.H. Kirchner and S.W. Zhang. (1991). “Range perception through apparent image speed in freely-flying honeybees”, Visual Neuroscience 6, pp. 519-535.

Srinivasan, M. V., S. W. Zhang, M. Lehrer and T. S. Collett. (1996). “Honeybee navigation en route to the goal: Visual flight control and odometry”, Journal of Experimental Biology 199, pp. 237-244.

Sull, S. and B. Sridhar. (1999). “Runway Obstacle Detection by Controlled Spatiotemporal Image Flow Disparity,” IEEE Transactions on Robotics and Automation, vol. 15, iss. 3, pp. 537-547.

Tang, L., S. Dian, G. Gu, K. Zhou, S. Wang and X. Feng. (2010). “A novel potential field method for obstacle avoidance and path planning of mobile robot,” IEEE International Conference on Computer Science and Information Technology (ICCSIT), vol. 9, pp. 633-637.

Tarutoko, Y., K. Kobayashi and K. Watanabe. (2006). “Topological Map Generation based on Delaunay Triangulation for Mobile Robot,” International Joint Conference SICE-ICASE, pp. 492-496.

Tchernykh, V., M. Beck and K. Janschek. (2006). “An Embedded Optical Flow Processor for Visual Navigation using Optical Correlator Technology,” in Proc. of the 2006 IEEE/RSJ Intl. Conf. on Intelligent Robots and Systems, Beijing, China, pp. 67-72

Thrun, S. (1998). “Learning metric-topological maps for indoor mobile robot navigation,” Artificial Intelligence, vol. 99, no. 1, pp. 21–71.

Thrun, S. (2000). *Probabilistic Algorithms in Robotics*, Technical report CMU-CS-00-126, Carnegie Mellon University.

Tian, T. Y., C. Tomasi and D. J. Heeger. (1996). “Comparison of approaches to egomotion computation,” in Proc. Computer Vision and Pattern Recognition, IEEE Computer Society Conference, pp. 315- 320.

Tsai, C., S. M Hsieh, Y. P. Hsu and Y. S. Wang. (2009). “Human-Robot Interaction of an Active Mobile Robotic Assistant in Intelligent Space Environments,” in IEEE International Conference on Systems, Man and Cybernetics, SMC 2009, pp. 1953-1958.

TUG (University of Maryland, www.umm.edu/news/releases/robot.html)

Tunstel, E. (1996). “Mobile Robot Autonomy via Hierarchical Fuzzy Behavior Control,” 6th Intl. Symp. on Robotics & Manuf., WAC'96, Montpellier, France, pp. 837-842

Wang, D., Y. Zhang and W. Si. (2011). “Behavior-based hierarchical fuzzy control for mobile robot navigation in dynamic environment,” Control and Decision Conference (CCDC), China, pp. 2419-2424.

Watanabe, M., K. Onoguchi, I. Kweon and Y. Kuno. (1992). "Architecture of behavior-based mobile robot in dynamic environment," in Proc. IEEE International Conference on Robotics and Automation, vol. 3, pp. 2711-2718.

Yung, N. H. C. and C. Ye. (1996). "An Intelligent Navigator for Mobile Vehicles", Proceedings of the International Conference on Neural Information Processing, Hong Kong, pp. 948-953.

Zhang, Z., Deriche, R., Faugeras, O. and Luong, Q. T. (1994). *A robust technique for matching two uncalibrated images through the recovery of the unknown epipolar geometry*, Research Report, RR-2273, Sophia-Antipolis: INRIA Sophia-Antipolis.

Zucchelli, M., J. Santos-Victor and H. I. Christensen. (2002). "Constrained structure and motion estimation from optical flow," in Proc. Pattern Recognition, 16th Intl. Conf., vol. 1, pp. 339 – 342.

Zufferey, J. and D. Floreano. (2006). "Fly-Inspired Visual Steering of an Ultralight Indoor Aircraft," IEEE Transactions on Robotics, vol. 22, No. 1, pp. 137-146.

APPENDIX A

Figures and data of the simulations

A.1. Parameters of the simulated environment

The simulated environment was generated by the rendering-program POV-Ray (version 3.6) enabling 3D environment simulation. The POV-Ray program reads information about objects and lighting in a scene stored in a text file and then generates an image of the scene from the view point of a camera described in the text file. The images can be generated in a very high quality with realistic reflections, shading, perspective and other effects.

The simulated 3D environment was constructed as an office scenario that includes four rooms R1 to R4, four room doors Dr1 to Dr4, three corridor doors Dc1 to Dc4, and four corridors Crd1 to Crd4, as illustrated in Fig. A.1. The sizes of the environment, including the walls, doors, corridors, and rooms, with the unit of length in centimeter (cm) are shown in Table Fig. A.1.

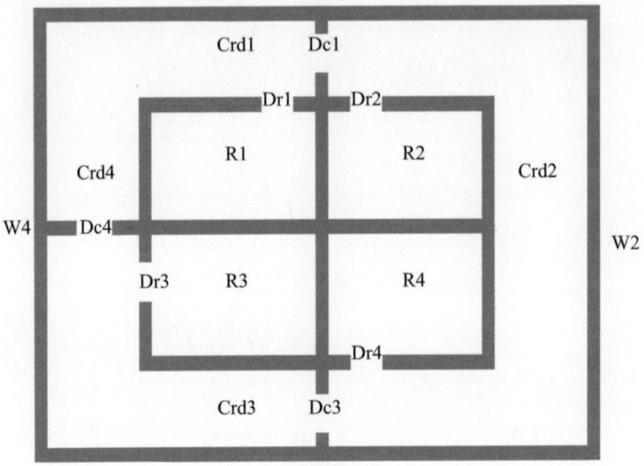

Fig. A.1. The simulated environmental model

Table A.1: Sizes of the simulated environment

Object *Wall*	Size in cm *high x wide x thick*	Object *Corridor*	Size in cm *long x wide*	Object *Door*	Size in cm *high x wide*	Object *Room*	Size in cm *long x wide*
W1	350 x 1990 x 30	Crd1	1930 x 290	Dc1	350 x 130	R1	600 x 415
W2	350 x 1560 x 30	Crd2	1500 x 320	Dc3	350 x 130	R2	600 x 415
W3	350 x 1990 x 30	Crd3	1930 x 290	Dc4	350 x 130	R3	600 x 415
W4	350 x 1560 x 30	Crd4	1500 x 320	Dr1	250 x 130	R4	600 x 415
				Dr2	250 x 130		
				Dr3	250 x 130		
				Dr4	250 x 130		

A.2. Parameters of the simulated furniture

The simulated furniture includes several chairs, given the symbols from C11 to C42, and five tables, given the symbols from T11 to T42. The furniture positions are shown in Fig. A2. The sizes and coordinates of the chairs and tables are displayed in Table A.2 and Table A.3, respectively.

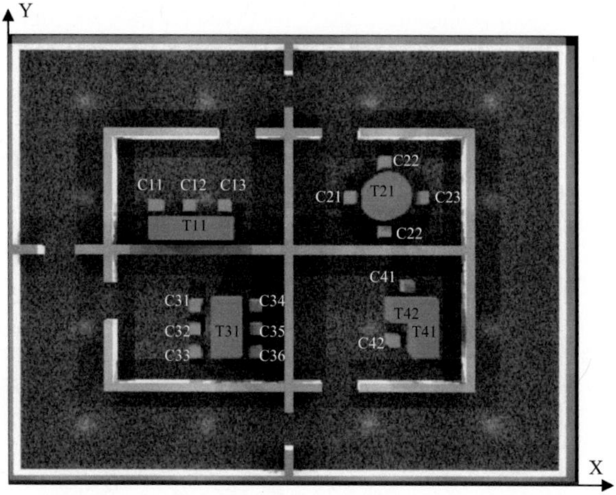

Fig. A.2. Furniture positions in the simulated environment

Table A.2: Sizes and coordinates of the simulation chairs

Furniture *Box chair*	Size in cm *wide x high*	Coordinate (X , Y)	Furniture *Box chair*	Size in cm *wide x high*	Coordinate (X , Y)
C11	40 x 60	(480, 950)	C31	40 x 60	(620, 600)
C12	40 x 60	(600, 950)	C32	40 x 60	(620, 520)
C13	40 x 60	(720, 950)	C33	40 x 60	(620, 440)
C21	40 x 60	(1160, 980)	C34	40 x 60	(830, 600)
C22	40 x 60	(1280, 1100)	C35	40 x 60	(830, 520)
C23	40 x 60	(1410, 980)	C36	40 x 60	(830, 440)
C24	40 x 60	(1160, 860)	C41	40 x 60	(1360, 570)
			C42	40 x 60	(1310, 480)

Table A.3: Sizes and coordinates of the simulation tables

Furniture *Rectangle table*	Size in cm *wide x long x high*	Coordinate (X , Y)	Furniture *Round table*	Size in cm *radius x high*	Coordinate (X , Y)
T11	80 x 280 x 100	(490, 850)	T21	80 x 100	(1300, 1000)
T31	100 x 200 x 100	(700, 450)			
T41	80 x 200 x 100	(1390, 450)			
T42	80 x 100 x 100	(1310, 570)			

A.3. Parameters of the static obstacles

The static obstacles are shaped like cylinder objects. These objects are randomly arranged along the corridors as illustrated in Fig. A.3. The sizes and coordinates of the static obstacles are listed in Table A.4.

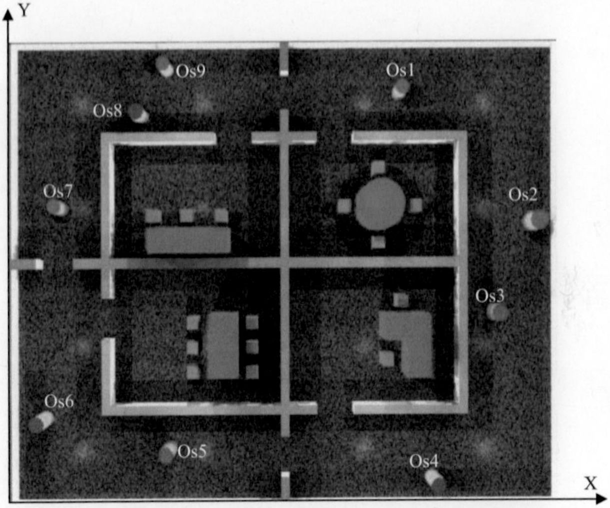

Fig. A.3. Positions of the static obstacles in the simulations SIM-s

Table A.4: Sizes and coordinates of the static obstacles

Static obstacle *Cylinder object*	**Size in cm** *high* x *radius*	**Coordinate** (X , Y)
Os1	70 x 25	(1380, 1380)
Os2	65 x 32	(1860, 950)
Os3	65 x 28	(1720, 650)
Os4	60 x 26	(1500, 200)
Os5	60 x 26	(550, 200)
Os6	60 x 26	(100, 300)
Os7	70 x 24	(160, 1000)
Os8	85 x 26	(450, 1300)
Os9	85 x 26	(540, 1460)

A.4. Parameters of the additional static obstacles

To generate the area with high density of static obstacles, a large number of static obstacles shaped like cylinder objects are randomly arranged as illustrated in Fig. A.4. The sizes and coordinates of the additional static obstacles are listed in Table A.5.

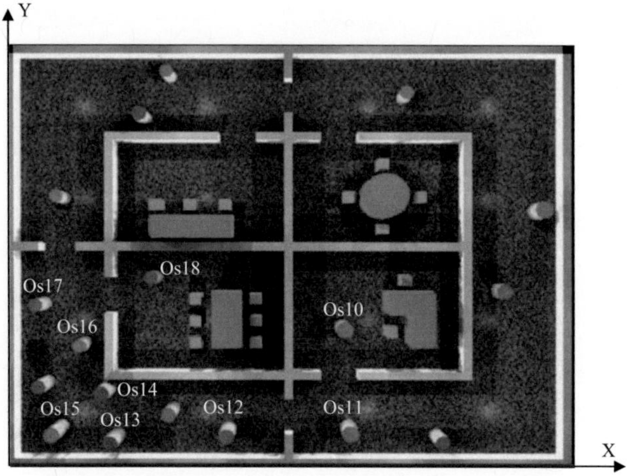

Fig. A. 4. Positions of the addition static obstacles in the simulations SIM-n

Table A.5: Sizes and coordinates of the additional static obstacles

Static obstacle *Cylinder object*	**Size in cm** *height* x *radius*	**Coordinate** (X , Y)
Os10	75 x 30	(1160, 510)
Os11	75 x 30	(1180, 130)
Os12	90 x 28	(750, 125)
Os13	90 x 28	(350, 100)
Os14	90 x 28	(320, 280)
Os15	90 x 28	(150, 130)
Os16	80 x 25	(240, 490)
Os17	80 x 25	(90, 600)
Os18	80 x 25	(490, 650)

A.5. Parameters of the dynamic obstacles

The dynamic obstacles are shaped like cone objects. The movement paths of these obstacles are recorded by the ground truth (in yellow) as illustrated in Fig. A.5. The sizes of the dynamic obstacles are shown in Table A.6.

Some dynamic obstacles move linearly, while others move casually. However, each of them moves around a bounded area. The coordinates of the bounded areas are illustrated in Table A.6.

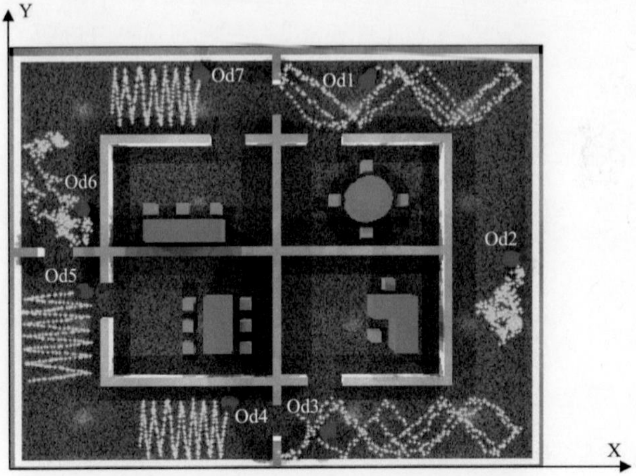

Fig. A.5. Paths of the dynamic obstacles in simulations SIM-d

Table A.6: Sizes and bound areas of the dynamic obstacles

Dynamic obstacle *Cone object*	**Size in cm** *height* x *top radius* x *base radius*	**Bound area** [Xmin, Xmax, Ymin, Ymax]
Od1	80 x 16 x 32	[990, 1910, 1250, 1470]
Od2	80 x 16 x 32	[1650, 1900, 310, 1200]
Od3	80 x 16 x 32	[990, 1910, 50, 270]
Od4	80 x 16 x 32	[320, 920, 50, 270]
Od5	80 x 16 x 32	[50, 310, 50, 750]
Od6	80 x 16 x 32	[50, 310, 820, 1470]
Od7	80 x 16 x 32	[320, 920, 1250, 1470]

A.6. Data-index of the simulation implementations

The objective of the simulation experiments is to test and verify the proposed algorithms and control system design with various navigation tasks in a closed-loop configuration. These simulations have been implemented in different scenarios and various operating conditions, for example, for coping with static obstacles and dynamic obstacles and for working in large and narrow areas.

The data indices of the implemented simulations in different scenarios of navigation tasks are listed in Table A.7. The first column of the table is the number of simulations. The second column shows the simulation data by the indices of the mat-files, of which the name means start node, end node, date of simulation implementation. The third column displays the navigation tasks. The last column informs the status of implemented simulation, where "success" means a navigation task was completed and "fail" means a navigation task was not completed.

Table A.7: Implemented simulations in different scenarios of navigation tasks.

No.	Index of mat-file	Navigation task and scenario	Status
01	N15_N03_20111216	Movement, corridors, doors, only furniture	success
02	N03_N10_20111217	Movement, corridors, doors, only furniture	success
03	N10_N15_20111218	Movement, corridors, doors, static obstacles	success
04	N15_N05_20111218	Movement, corridors, doors, static obstacles	success
05	N17_N19_20111219	Movement, corridors, doors, rooms, high density of static obstacles	success
06	N19_N17_20111219	Movement, corridors, doors, rooms, high density of static obstacles	success
07	N02_N11_20111222	Movement, corridors, doors, dynamic obstacles, where $\max(\mathbf{v}_{DynamicObstacle}) \approx 60\% \max(\mathbf{v}_{Robot})$	success
08	N09_N02_20111223	Movement, corridors, doors, dynamic obstacles, where $\max(\mathbf{v}_{DynamicObstacle}) \approx 80\% \max(\mathbf{v}_{Robot})$	success
09	N08_N02_20111222	Movement, corridors, doors, dynamic obstacles , where $\max(\mathbf{v}_{DynamicObstacle}) \approx 100\% \max(\mathbf{v}_{Robot})$	success
10	N13_N10_20120509	Movement, corridors, doors, high density of static obstacles	success
11	N07_N16_20120511	Movement, corridors, doors, dynamic obstacles, where $\max(\mathbf{v}_{DynamicObstacle}) \approx 120\% \max(\mathbf{v}_{Robot})$	fail
12	N13_N01_20120511	Movement, corridors, doors, static obstacle, and dynamic obstacles, where $\max(\mathbf{v}_{DynamicObstacle}) \approx 160\% \max(\mathbf{v}_{Robot})$	fail